亚太豪宅大赏

ASIA-PACIFIC MANSION COLLECTION

风靡大陆
SWEEPING MAINLAND

深圳视界文化传播有限公司 编

中国林业出版社
China Forestry Publishing House

序言 / PREFACE

风雨中方显设计本真
THE ESSENCE OF DESIGN LIES IN TRIALS AND HARDSHIPS

文/杜柏均
柏仁国际联合设计

Script/ Du Bojun
PRID International Design

从事设计已经近30年了,深深体会到设计行业辛苦与荣耀的错综复杂,但不得不肯定它使得我的生活日渐丰富。这是一个不容易的行业,在此向广大的设计师同行们致敬,有你们,人们的生活过的更有尊严与质量。

12年前我毅然放下台北的公司,只身到上海打拼,赶巧遇到上海的蓬勃发展时期,让我有了很好的业务机会。从汤臣高尔夫开始到西郊庄园,让我重新认识了豪宅的定义,也帮助了许多开发商从室内与生活的角度去定位了高级别墅的规范与标准。我在开发商与建筑师讨论过程中发现,开发商很多时候只能将硬件定义在人类生理的功能,精神上和心理上的功能还是需要室内设计师来细化,不是他们忽略了,而是这项工作没有明确的标准,且工作量巨大,所以说设计师的存在价值是非常有必要的。这也是导致许多设计师没有明确自己的价值而随波逐流,最终只能当甲方的绘图员。

经历了我国30年来的改革开放,人们对"住"的概念在不同阶段有着不同的要求,从刚开始的求有到求好,再到求精神层面上的满足,每阶段上都带有不同的意义。而我们国内设计的前沿梯队也已经进化到精神与尊严的阶段,到这时候已经不是简单的生理需求层面,项目的标准都需拉高到历史定位上才能给予甲方更大的满足。比如,豪宅要考虑的就包括家族精神的传承,而不只是考虑几房几厅几个卫生间的问题,而这也是我们设计师正在努力追寻的方向。

今有幸受邀写序,我不想说一些高尚的词句,只想与同业分享我的感受,在此国内外经济调整的时期也是设计师们重新思考自己的定位时期,审视自己惟有从原点出发,戒除拿来主义,不能一味追求公司做大,但求做精做实,让自己对得起项目。当你摸着心将项目当做自己家来做时,那恭喜你,你将在风起云涌时发光发热,若不是那么赶紧调整心态,恐怕将会在这波调控中湮灭。这个标准如何来判断,其时也很简单,回头翻阅你的业务来源,是业务开发而来还是回头客或口碑?所以诚实的面对自己,就能找出自我的优缺点,因而也就能找出定位与出路。如果每个设计师都有这样的想法,那中国设计师走向世界就不远矣。再次与大家共勉之!

I've been a designer for almost 30 years and realized deeply the complex feeling of glories and pains in this industry which has made my life more abundant gradually. This is an industry which is not easy, thus I should pay my respects to my fellow designers here that people live more honorable and qualified due to you all.

I left from the company in Taipei 12 years ago and worked alone in Shanghai when I met the booming period of Shanghai that brought me many good opportunities. From the projects in Tomson Golf Villa to Forest Manor, I recognized the definition of mansion for a second time as well as helping many developers to orient the standards of high-class villas from both the interior and living aspect. During the discussion between developers and architects, I find that developers could only define the conditions and facilities on the physiological functions of human beings many times, while interior designers are needed to detail them on spiritual and psychological functions since there are not any specific standards with huge workload yet not their neglect. As a result, it is quite necessary of the value of designers. However, those who have not realized their value and drifted with the current could only be a draftsman at last.

People have different demands of "dwelling" in different periods after the reform and opening-up policy for 30 years. Spaces for living, good spaces for living and then the spiritual satisfactions with different meanings in different periods. While the leading design teams in our country have reached the period of spirit and dignity, so that designs are no longer related with simple physiological needs, while the standards of projects must be raised to a historical level to give more satisfaction for the buyers. For example, the inheritance of family spirit should be considered in a mansion, yet not simply the numbers of rooms, halls or bathrooms, which is the orientation that we designers are pursuing for.

It's my honor to be invited to write a preface, thus I don't want to use a lot of lofty words but my own feelings. During the period of domestic and international economic adjustments, we should rethink our orientation. Starting from the origin to take self-examinations, we should get rid of the Copinism and aim at elaboration and honesty yet not merely the scale of the company. If you treat each project as your home, then congratulations that you will shine in the market, and if not, you have to adjust your attitude or you will be annihilated in this regulation and control. How to judge the standard is easy actually, just look over on your sources of business, new developments or returned customer and public praise. As long as you face to yourself honestly, you could find your merit and demerit as well as the orientation and solutions. Chinese designers would go to the world in the near future if each designer has the same idea. Share with you all.

目录 / CONTENTS

风靡大陆
SWEEPING MAINLAND

本系列图书主要汇聚亚太地区顶尖设计人才至新至TOP的设计，希冀通过视界出版的努力，为国内外精英设计师构筑国际化交流平台，打造精英设计师品牌。

This series of books mainly bring together the top design talents to new TOP designs of the region, hoping that with the efforts of Design Vision, these books can help to build an international communication platform for excellent designers at home and abroad, thus creating brands of elite designers.

凌 子 达

大道至简 清新儒雅

CONCISE, FRESH AND REFINED

008

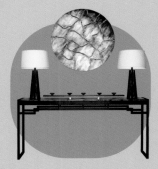

LSDCASA

再续东方美学

THE EXTENSION OF EASTERN AESTHETICS

022

连 志 明

轻奢法式新古典

NEO-CLASSICAL FRENCH STYLE WITH SLIGHT LUXURY

038

连 志 明

迷醉的法式空间

THE ENCHANTING FRENCH SPACE

060

柏 舍 励 创

静雅居

A QUIET AND ELEGANT RESIDENCE

076

柏 舍 励 创

融会

INTEGRATION

086

杜 柏 均

270度视角俯瞰都市繁华

270° VIEW OF URBAN PROSPERITY

100

毛明镜、昌影

感怀静时光

REMINISCENCE OF THE TRANQUIL TIME

110

毛明镜、昌影

跨界时尚

CROSS-BOUNDARY FASHION

118

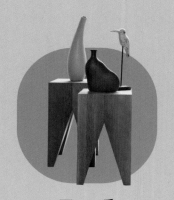

王 兵
来自北欧的清风
REFRESHING BREEZE FROM
NORTHERN EUROPE
124

王 兵
意式柔情 多彩之歌
ITALIAN TENDERNESS WITH
COLORFUL SONG OF PRAISE
138

彭 征
壹城壹墅
ONE CITY ONE VILLA
150

壹挚设计集团
摩登雅痞
MODERN YUPPIE
164

壹挚设计集团
摩登东方
THE MODERN ORIENT
176

谢 辉
圆舞曲
A WALTZ
186

全筑装饰
依景造城 共生共荣
BUILT ALONG THE SCENERIES
WITH CO-EXISTENCE AND COMMON
PROSPERITY
198

DAOKEY
轻蓝主义
SLIGHT BLUE-ISM
206

李孙设计
法式情缘
LOVE OF FRENCH STYLE
230

顾惠娟

复古英伦风

VINTAGE BRITISH STYLE

246

何文哲、唐瑶华

摩登轻法式

SLIGHT FRENCH STYLE IN FASHION

262

洪德成

融入"历史与文化"中的低调奢华

LOW-KEY LUXURY BLENDED IN "HISTORY AND CULTURE"

276

岳 蒙

橄榄树之恋

LOVE OF THE OLIVE

290

何永明

绿野仙踪

THE WIZARD OF OZ

296

Rich

日暖绿城·典藏美宅

GREENTOWN WITH WARMTH·A MEMORABLE RESIDENCE

306

大道至简 清新儒雅
CONCISE, FRESH AND REFINED

项目名称 ｜ 中信泰富别墅样板房
设计公司 ｜ KLID 达观建筑工程事务所
设 计 师 ｜ 凌子达
项目地点 ｜ 上海
项目面积 ｜ 600 ㎡
主要材料 ｜ 大理石、木板等
摄 影 师 ｜ 上海达观建筑工程事务所

KLID 达观建筑工程事务所 创始人

凌子达，1973年出生于台湾高，1999年毕业于台湾逢甲大学建筑系。2001年在上海成立"达观国际建筑室内设计事务所"，致力于建筑室内空间设计领域。2006年出版个人作品集《达观视界》。2009年取得法国国立工艺学院（CNAM）建筑管理硕士学位。

坚持"达者为新，观之有道"的设计宗旨，用豁达的胸襟看世界，以开阔的视野带来新的创意和想法。2014年荣获德国红点设计大奖之红点奖，以及荣获德意志联邦共和国国家设计奖，还有英国BLUEPRINT设计大奖等。2015年德国设计大奖，德国红点奖特别提名，美国ARCHITIZER A+ AWARDS最佳荣誉奖以及美国IDEA室内设计大赛优秀奖，同时获得新加坡SG Mark设计大奖，英国aisa proerty awards最高荣誉奖，还有中国金盘奖最佳年度预售楼盘奖等。

此次设计仍旧延续了新东方的基调。凡文人雅士都想拥有一间雅致别墅、享受人生成功的乐趣。

整体住宅的设计以暖色为主调、用柚木和大理石做饰面,将中国的造型概念如:虚与实、藏与露、借景对景等手法结合现代元素,运用在设计中。

It continues neo-oriental tone again in this case. Refined scholars are all willing to have an elegant villa to enjoy the pleasure of the success in life.

The overall design is dominated by warm colors, while teakwood and marble are used as the veneer. The designer combined many techniques in the modeling concept of Chinese style with modern elements in this design, such as the true and false, the clarity and obscurity, as well as borrowing scenery.

古人主张以白反映内心世界，避免俗气装饰。整体空间用大量米色做饰面，衬托陈列品。并使用大型木屏风等间隔物，将中国文人装修所主张的"儒"的思想，通过装修虚与实、藏与露、借景对景等手法体现出来。窗户要透光通风，借景为画的审美作用。装修中的桌椅陶瓷字画皆出自名师上乘之作。楼梯口鸟语花香的设计，让人沉浸在自然地感觉中，上下楼梯时也别有不一般的风味。

设计师在墙面适当留白，给主人挂画，并通过灯光设计，使墙面具有更强的观赏性。半通透的屏风，保持空间良好的通风和透光率，营造出"隔而不离"的效果，让家庭空间进行有效的互动，却又不会使空间过于封闭；屏风以简单的线条为主体，材料选择了柚木，看上去不失中式的韵味，又能表达现代的个性。它为整个别墅提供了充足的光线与丰富的空间层次。

The ancients advocated white color to reflect the inward world and avoid the vulgar decorations. Large amount of beige color is used on the veneer to embellish the exhibitions. Large wooden screens and other partitions are also adopted to show the thought of "Confucianism" that advocated by Chinese scholars through techniques. The windows should be ventilated with light, achieving the aesthetic function. The tables, chairs, porcelains and drawings all come from the wonderful works of famous masters. The design with birds and flowers at the staircase makes people feel as if they were standing in nature, bringing a unique flavor.

The designer left several empty spaces on the wall for the owner to hang on drawings, making the wall surface have the ornamental value better through the lighting design. The semi-transparent screen kept a good ventilation and luminousness in the space, creating an effect of "partition yet connected", allowing the effective interaction in the family space yet not making it enclosed. The screen features simple lines and teakwood, showing the Chinese flavor as well as modern individuality, providing sufficient light and rich layers for the whole villa.

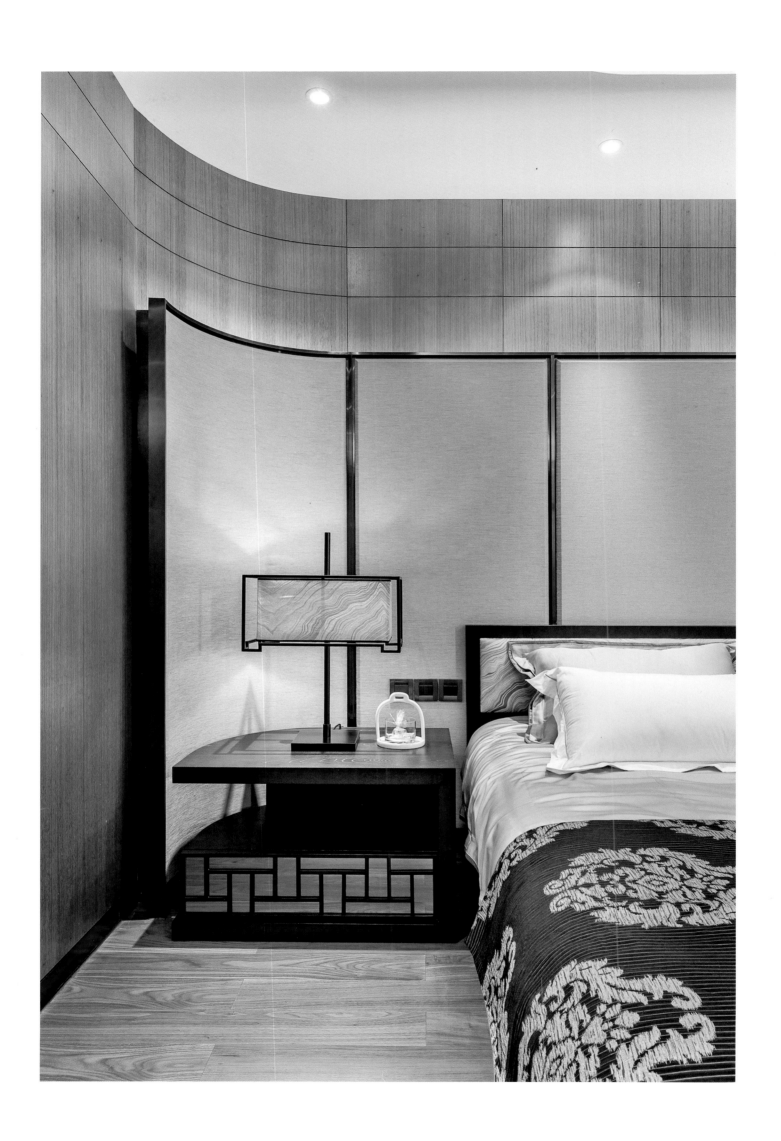

再续东方美学
THE EXTENSION OF EASTERN AESTHETICS

项目名称 ｜上海绿地海珀佘山别墅
软装设计 ｜LSDcasa
项目地点 ｜上海
项目面积 ｜270 ㎡

LSD®casa

LSDCASA

　　LSDCASA隶属于深圳市进化生活家居用品有限公司，由葛亚曦先生创立。在深圳、北京等地设有分公司及办事处。面向国内外开发商及私人客户，针对酒店、大型商业空间、会所、别墅、示范单位等，提供软装设计及顶级定制服务。此外，依托强大的家居用品及艺术品采购渠道及网络，在世界范围内搭建顶级家居品牌的整合平台。

　　立场是战斗性的，但设计师的战斗方式，必须是极致的美。LSDCASA在上海绿地海珀佘山别墅这个作品中，从文化根源追溯设计根本。宋代的美学，日式的工匠精神，现代中国人的生活研究，这些都是设计之前必做的功课。

　　在这里，设计师将对"天人合一"的意境诉求融汇于现代生活情境之内。设计保留传统中式风格含蓄秀美的设计精髓之外，呈现现代、简约、秀逸的空间，使环境和心灵都达到灵与静的唯美境界，迸发出更多可能性的联想。化繁为简，吐故纳新是该居所的创作内核。设计放弃对风格样式的表象追求，在情绪、文化、气质、认同层面，寻找可以联系、沟通、协调的路径，以此表达人的精神诉求。

　　客厅与餐厅，一气呵成，极简示人。餐厅吊灯的形式、材质与空间背景相呼应，铅白色的桃花在玄青花器上显得分外典雅，秋色的漆画在昭示着相近文化背后的绚丽山景。

　　配合空间，物件主要以哑光釉面和陶面的器皿为主，与之对应的陶瓷器皿与漆器器具，亦是源自中国传统皇宫家具和日本传统漆品的样式体现，而局部则以点缀加入商夏青铜古董摆件增加空间的收藏历史意味。

A standpoint is a battle, while the designer's way of a battle must be the extreme beauty. In this project, LSDCASA traced the essence of design from the source of culture, such as the aesthetics in the Song dynasty, the spirit of Japanese craftsman and modern Chinese lives, which are the necessary things before designing.

Designers blended the flavor of "syncretism between heaven and human" into modern life. This design remained the implicit and graceful design essence of traditional Chinese style and presented a modern, concise and elegant space to reach the wonderful realm with intelligence and tranquility both on the environment and the mind, thus more imaginations will be produced. Making hard things simple as well as getting rid of the stale and taking in the fresh is the core of creation in this residence. This design focuses on the aspects of emotion, culture, temperament and recognition to find the way to be connected, negotiated and coordinated instead of superficial pursuit on styles, expressing the spiritual appeal of human beings.

The living room and dinning room were connected with conciseness. The form and materials of the droplight in the dinning room echoed with the background of the space, the white peach blossom on the utensil in blue and white appeared quite elegant, while the painting showed the gorgeous landscape of similar cultures.

Utensils with lusterless glaze and ceramic glaze were dominated in the space, the porcelain and lacquerware related were derived from the patterns of traditional furniture in Chinese palace and Japanese traditional lacquer, while the antique bronze ornaments like those in the Xia and Shang dynasties added the flavor of collection and history into the space.

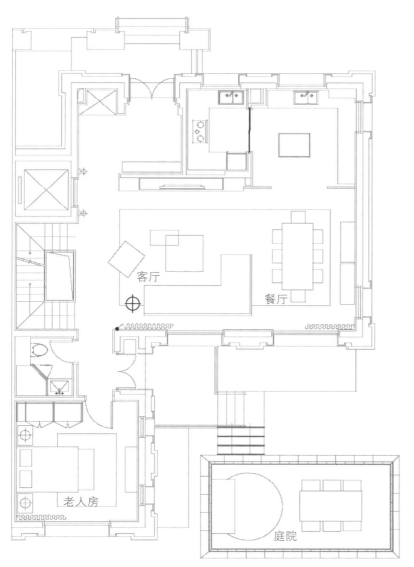

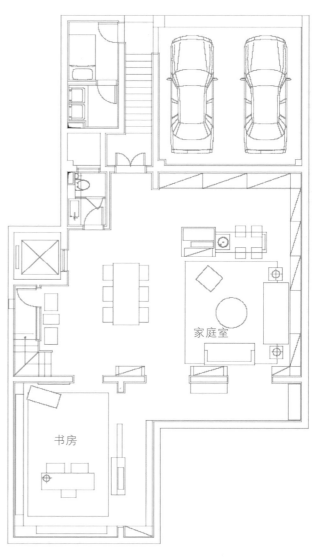

此外、桃花、梅花等传统花艺的介入，在空间延伸了无限的意境，插花的方式更是讲求节奏与韵律感，以"象外之意，景外之象"，"韵外之致，味外之旨"诠释空间的文化精神。

直接而干练的线条，自由放松的尺度，是对应空间最贴切的诠释。茶白的清雅自然、赭石的稳重硬朗、靛蓝的深邃蜿蜒，犹如在山水之间，意境之中。棉麻质感的材质舒适而温婉、编织的茶几是东方印象的细致趣味，加以金属铜色的轻微点缀，让空间在质朴雅致的意境中又份外的提炼出一丝丝的当代气质与空间契合，不温不燥、不多不少、恰当刚好。

负一层家庭房与书房相连，空间色调除了深色的木作色之外，以蓝色、墨绿色、灰色为主，用色不多却非常讲究，这些来自大自然的颜色，在表现含蓄的同时，也带来视觉上的一抹清新。

私密的卧室沿袭整个空间的清雅与平和，空间色彩之间、形式之内无一不在提示着过去、现在与未来，东方人文的演变与糅合。

The introduction of peach blossom, plum blossom and other traditional floriculture extended the space with infinite meaning. The methods of floriculture focused on rhythm, interpreting the cultural spirit of the space.

Direct and neat lines as well as free and eased scale are the most proper interpretation of the space. The elegance and nature of white, the steadiness and toughness of ochre and the depth and wriggle of indigotin seemed as if they were in the landscape. The comfortable and gentle cotton and linen, the knitted tea table with delicacy and interest of oriental impression and the little embellishment of metallic copper brought a modern temperament within the pristine and elegant atmosphere, which were appropriate in the space.

The family room and study were connected in the basement, the color tone of which was dominated by blue, dark green and gray besides the dark wooden color. These delicate colors from nature brought a little fresh sense as well as connotation.

The private bedroom followed the elegance and peace in the overall space, where the colors and forms all showed the transformation and integration of the past, the present and the future as well as oriental culture.

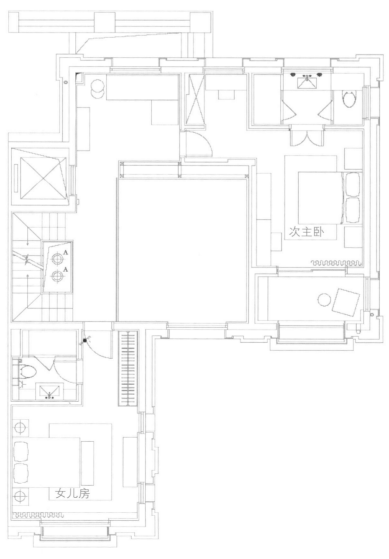

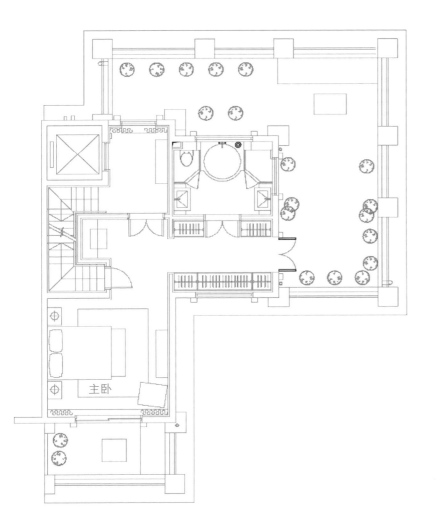

连志明 / ZHIMING LIAN

北京意地筑作室内建筑设计事务所 创始人兼设计总监

大然设计品牌 创始人

  连志明，北京意地筑作室内建筑设计事务所及大然设计品牌创始人，中央美院客座教授，清华美院艺术陈设设计研修班讲师。自1995年至今，连志明在室内建筑空间设计的领域里已经深耕20载，其设计作品曾荣获国内外多项奖项。作为一名设计师，他将自己对于空间的认识和理解倾注在室内环境中的空间形态、光影，以室内材料肌理的研究上。同时，由于早年在法国巴黎高等室内建筑及广告设计艺术学院（ESAG Penninghen）的教育经历，连志明对于法式生活方式的把握，以及空间之中精神层面的当代性认识，有着他独到之处。他善于在室内环境中营造出法式新古典主义的浪漫与奢华风情，亦能够以全新视角演绎传统文化在设计领域的当代表达，并兼具软装设计师的细腻与建筑师对于空间关系的理解。此外，出于对产品设计的热爱，连志明创建的大然设计家具产品品牌亦获得斐然成绩，其产品行销全球30个国家并饱受好评。

轻奢法式新古典
NEO-CLASSICAL FRENCH STYLE WITH SLIGHT LUXURY

迷醉的法式空间
THE ENCHANTING FRENCH SPACE

轻奢法式新古典
NEO-CLASSICAL FRENCH STYLE WITH SLIGHT LUXURY

受到18世纪新古典主义的影响，现代法式设计中保留了来自文艺复兴时期以来欧洲传统的视觉元素，纺织品上繁复的卷草纹样与木饰板上浅浮雕般的线条轮廓，勾勒出一种"欧罗巴印象"。然而，相较之于更早时期的装饰习惯，新古典风格却又减少了更多细枝末节的修饰，从而强调了整体的线条与轮廓。

Influenced by the neo-classicism of the 18th century, modern French designs have remained traditional visual elements since the Renaissance Period, such as the complex pattern of rolling grass on the fabric and the lines like bas-relief on the wooden veneer, outlining a kind of "Europa impression". However, compared with the decorative customs in earlier stage, neo-classical style reduces more minor details so as to enhance the overall lines and profiles.

设计公司	北京意地筑作装饰设计有限公司
设 计 师	连志明、张伟、徐辉
项目地点	北京
项目面积	820 ㎡
主要材料	大理石、窗帘、木作、镜子、壁纸等

客厅内繁复的地毯与墙纸，被纯净的白色立面分割开，从而避免了过度装饰带给人们的不安感受。舒缓的白色，也将空间形态更加完整清晰地描绘出来。而深色的铁艺和线角，则与之形成强烈的对比，并且让环境变得更加明亮和醒目——相对于古典巴洛克风格深浮雕与繁杂装饰带来的浓郁与阴沉，新古典主义显得更加柔和舒服。

就算不去仔细辨别，我们也可以看到，当代法式风格对于色彩和明暗的处理，也有着独特的印记。受益于19世纪法国印象主义对色彩理论的推进，法式风格中空间色调更加倾向于明快的色相对比，从而营造出甜美风情。与同样在19世纪诞生的美式殖民地风格不同，法式设计并不依赖于厚重的色调与皮革、高档木材等稳定质感所产生的奢靡，相反大量运用羊毛、真丝等等更为精巧的材质来呈现对富足生活的描绘。而这与法国人一直以来所标榜的生活美学息息相关。

The complex carpet and wallpapers are divided by the pure white facade in the living room to avoid the uneasiness brought by the excessive decoration. The soothing white depicts the space more completely and clearly. While the dark iron art and lines form a strong comparison, making the environment appear bright and eye-catching—compared with the heaviness and darkness of classical Baroque, neo-classicism presents more soft and comfort.

We could see that modern French style has its unique features on the handling of colors and shading. Benefited from the advance on color theory of French impressionism in the 19th century, the color tone in French style tends to be compared with bright colors in order to create a sweet and beautiful flavor. Different from the American colonial style at the same period, French designs are adopted with large amount of wool, silk and other delicate materials to present the luxurious life instead of heavy tone, leather, qualified wood and other extravagant materials, which is close to the aesthetics of life belong to the French.

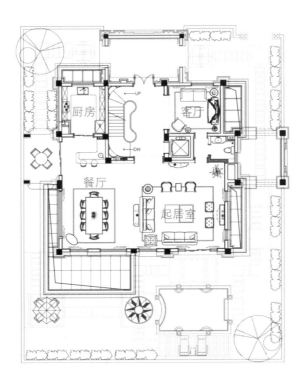

亚太豪宅大赏 III · 风靡大陆 | 045

在软装陈设中，流苏、蕾丝、刺绣让法式空间获得了更多的表现形式……得益洛可可时代欧洲对装饰美学的拓展，法式空间所涉及的装饰品类极其丰富。但同时，这也意味着对于设计师来说手中供货商名单是一个长长的列表，供应链的狭促将大大折损室内空间的品位与气度。此时此刻，现代生活方式与节奏，也令家具与软装产生了有别于一百年前的全新需要。

On the soft decoration, tassels, laces and embroideries make the French space achieve more patterns of manifestation...Benefited from the expansion of Rococo era on decorative aesthetics, French spaces are referred to various kinds of decorations. However, this means there will be a quite long list from the suppliers, so that the limitation of the supply chain reduces the taste and presence of the indoor space greatly. At the same time, modern lifestyles and rhythms produce new demands on furniture and soft decoration which are totally different from those in a hundred years ago.

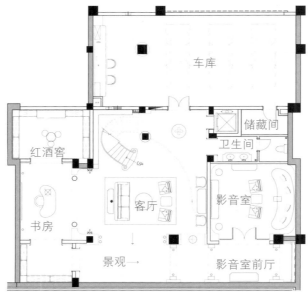

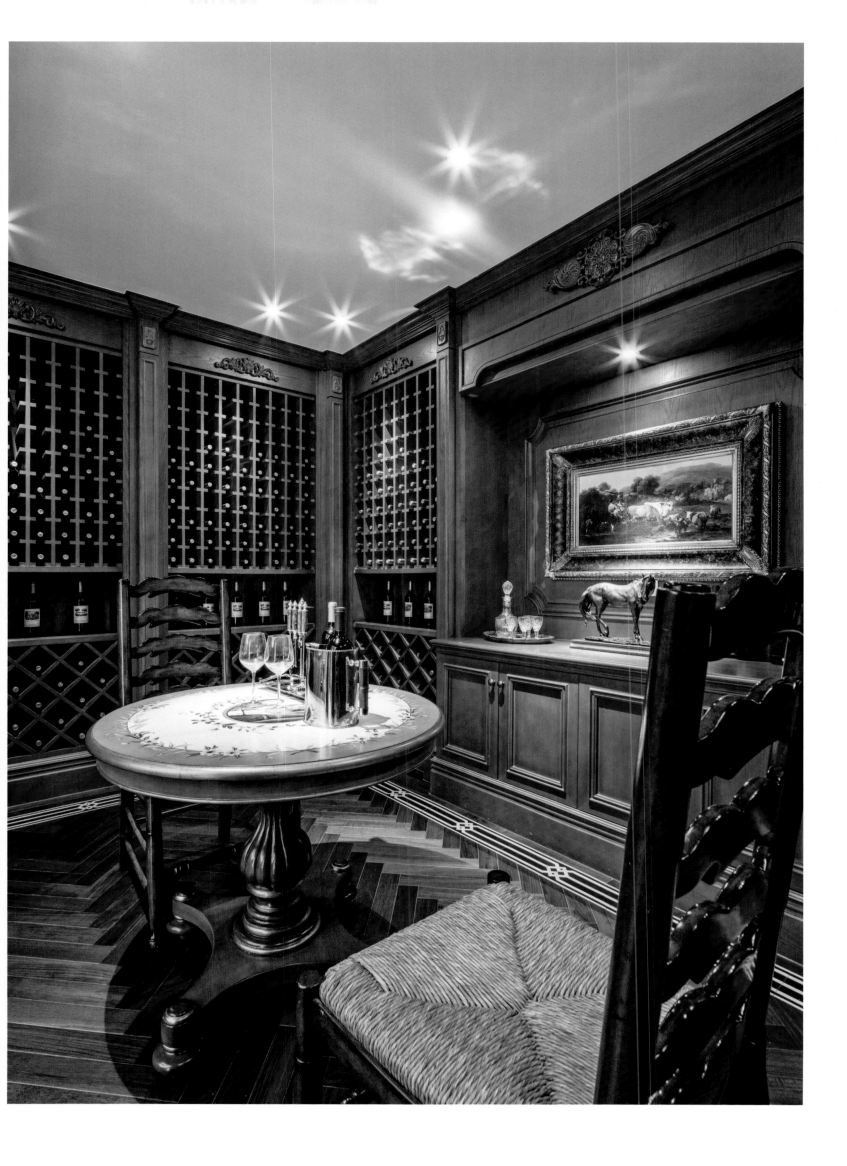

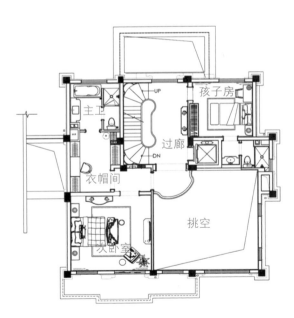

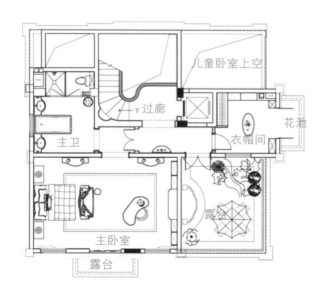

迷醉的法式空间
THE ENCHANTING FRENCH SPACE

设计公司	北京意地筑作装饰设计有限公司
设 计 师	连志明、张伟、徐辉
项目地点	北京
项目面积	584 ㎡
主要材料	大理石、窗帘、白色墙板、镜子、壁纸等

　　两层地下室空间的设计在补充楼上空间功能需求之外，又满足了业主的娱乐、收藏及展示需求。首先地下一层空间里运用采光井的设置，令地下空间增添自然光线与空气流通；其次，台球厅的设置，增加了家庭空间的娱乐性；大面积的多功能区域，为业主提供了宽敞的艺术品或设计品的收藏与展示空间。地下二层除了设置有必不可少的车库外，还有家庭放映室、读书室、酒窖、储物间、收藏室，真可谓一应俱全。

　　除此之外，最为吸引人眼球的即是空间的艺术品搭配。不管是具有法式风情的人物、风景画、还是趣味横生的现代动物主题，抑或是古典的乾隆狩猎图，以及国画图案的墙纸，如此"混搭"的艺术品都完美地融合在这个法式的空间中，并无疑成为了空间的点睛之笔。正如艺术本无国界一样，法式风格穿越时光的隧道，在现代的异域同样绽放着引人注目的花朵。

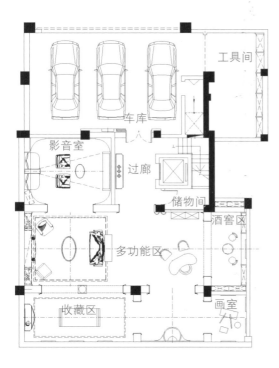

Two floors underground satisfy the owner's demands of entertainment, collection and exhibition besides supplementing the functions. The first floor underground is settled with a light well, adding natural light and ventilation into the underground space; the settlement of the billiard room adds the entertainment in the family space; large area of the multifunction space offers a spacious place for collecting and displaying the art works and designs. The second floor underground houses a projection room, a reading room, a cellar, a storage and a collection room besides the necessary garage.

In addition, the collocation of art works among the spaces is eye-catching, such as the paintings with French flavor, the interesting theme of modern animals, the classical drawing of Emperor Qianlong's hunting, and the wallpapers with national patterns, the mix of different art works of which are combined perfectly in French style, becoming the highlights. Just as there's no boundary of art, French style is blooming in foreign lands now through the tunnel of times.

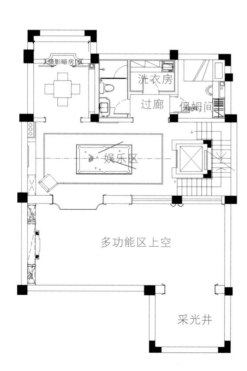
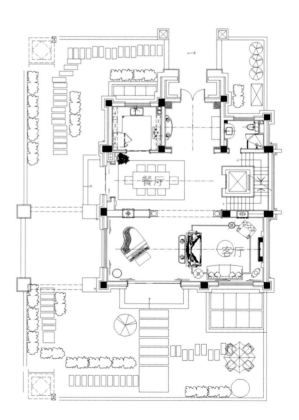

　　本案打造的是法式经典空间，除却空间透露的时尚优雅的气息外，让人眼前一亮的还有"混搭"的艺术品陈设，以及空间高贵而又不失理性的柔美线条。在这浪漫、奢华的法式情调里，舒适感、艺术感同样没有缺席。为了满足业主社交与多种娱乐活动方面的需求，设计师在别墅空间功能的划分上可谓独具匠心，首先一层设置有宽敞的餐厅，起居室的旁边设有开放的钢琴演奏区，可供小型聚会或者家庭聚会使用。二层空间布置有男女主人的卧室及女儿的卧房。三层空间设有一间客房、一处衣帽间以及一块供闲聊聚会的区域。

This case aims at creating a classical space in French style, which features the "mixed" display of art works and the noble and reasonable lines besides the fashionable and elegant flavor. It appears comfortable and artistic within this romantic and luxurious French flavor. In order to satisfy the owner's demands of social contact and multiple entertainments, designers divide the spaces ingeniously. The first floor houses a spacious dinning room, a living room and an open piano playing area beside which could be used to hold small parties and family gatherings. The second floor houses the master bedroom and the daughter's room. The third floor houses a guest bedroom, a cloakroom and an area for chatting and gathering.

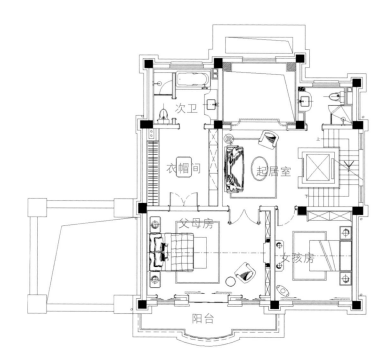

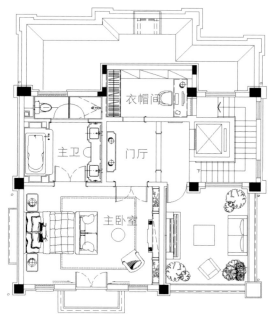

在这个具有584平方米的空间中,经典的法式风格充斥其中,不管是大面积淡灰白色、咖啡色、灰蓝色、米色的使用,还是米黄石材、高级灰石材、拼花木地板、黑胡桃木饰面、簇花壁纸的选择,抑或是深咖啡色封闭漆、蓄绒及织染丝绸布面家具的摆放,再加上铜蜡烛水晶吊灯、布艺帘、法式饰品、古董的点缀,无疑让人在这古典而又时尚,浪漫而又奢华的空间里迷醉。秉持典型的法式风格搭配原则,餐桌和餐椅尽显高贵。花鸟的图案下,镶有金色雕花。客厅中扶手椅椅腿的弧形曲度尽显优雅矜贵,而在白色的卷草纹窗帘、水晶吊灯、落地灯、瓶插洋蓟花的搭配下,浪漫清新之感扑面而来。在卧室中,线条的繁复以及色彩的透明浅淡,都体现出法式风格的内涵。花朵图案的墙纸、地毯,以及具有法式线条的花瓶摆设,都传递着法式的精华,让人耳目一新。

The classical French style is dominated throughout this space with 584 square meters. No matter the use of large area of light offwhite, coffee, gray blue and beige, the choice of cream-colored stone, advanced gray stone, parquet wood floor, black walnut veneer and wallpapers, or the settlement of dark coffee paint, velvet, furniture with silk cover, added with the embellishment of chandeliers, curtains, French accessories and antiques, all make people feel intoxicated in this classical, fashionable, romantic and luxurious space. The dining table and chairs appear noble since they are adopted with the collocation of French style. The patterns of flowers and birds are inlaid with gold carvings. The legs of the armchair in the living room appear its elegance and nobility, which present a romantic and fresh flavor through the collocation of the white curtain, chandeliers, floor lamps and the artichoke in the vase. The complex lines and colors with different shades in the bedrooms show the connotation of French style, where the wallpaper and carpet with patterns of flowers, and the vases with French lines convey the essence of French style, making people feel fresh and new.

柏舍励创 PERCEPTRON

柏舍励创设计机构致力于高端设计品牌的革新与实践，秉承鼓励创新的理念，合理分配资源，推动多元化艺术形式的发展，积极应对市场挑战，为设计团队构筑坚实的设计与管理体系，提供综合的服务平台。

机构旗下设有四家专属的室内设计品牌：柏舍设计拥有完善的研发及实施系统；本则创意推崇在意境中探索精神世界；5+2设计善于发掘生活艺术点滴；利奥软装设计具备完整资源体系，主张用陈设对话艺术。

静雅居
A QUIET AND ELEGANT RESIDENCE

融会
INTEGRATION

静雅居
A QUIET AND ELEGANT RESIDENCE

项目名称 ｜星星广场四期18#样板房
设计公司 ｜柏舍设计（柏舍励创专属机构）
项目地点 ｜广东佛山
项目面积 ｜320 ㎡
主要材料 ｜英伦玉石、佛罗伦棕、皮革、玫瑰金拉丝钢等

　　中国风正在全世界范围内大行其道甚至影响着设计潮流，作为本土设计师应该认识到，中国元素并不是"符号式"的表达，要让中国元素的使用不肤浅地流于表面，而是有机融入到产品当中，真正提升整个产品的品质、从而让产品变得更接地气。

Chinese style is popular all over the world which even has an influence on the trend of design. As local designers, we should realize that Chinese elements are not the expression of "symbols", and what should we do to blend them into the products to enhance the quality and make the products related to ordinary people instead of being used superficially.

佛山是一座历史悠久的文化名城，明清时代就被列为中国古代四大名镇之一，拥有浓厚的岭南文化气息。技巧精湛的民间艺术剪纸和陶艺仍有迹可寻，这些古老的文化方式正在以另一种形式存在于人们的生活中，慢慢发迹、延续。

设计师提取山水元素，水水潋潋，山岛竦峙。背景墙面由玫瑰金钢组成"山峦层叠"，地毯则提取水为元素，山水呼应，碧波荡漾，营造一个意境悠远、高雅大气的空间氛围。淡淡的来，淡淡的去，淡淡的相处，给人以宁静，予己以清幽；静静的来，静静的去，静静的守望，给人以宽松，予己以从容。当我们的生活累时，迷茫时，请停留在我们各自的心灵憩息之所，浮华便已成云烟，喧嚣也已不再，只和心静静的对白，只将心轻轻的靠岸……

Foshan is a famous cultural city with a long history, which was listed as one of the four famous towns of ancient China in the Ming and Qing dynasties, and covers a profound culture of Lingnan. The exquisite and skilled folk art scissor-cut and pottery could be traced until now, while these old cultural forms are existed, developed slowly and continued in life with another form.

The designer selected landscape elements to present a scene of water waves. The background wall was formed like "a stretch of mountains" made of rose gold steel, while the carpet was adopted with the element of water, creating an elegant and generous atmosphere with a capacious artistic conception through the echoing of mountain and water. Coming, leaving, and getting along lightly, gives others tranquility and brings freshness for yourself; Coming, leaving and waiting silently, gives others relaxation and brings leisure for yourself. When we are tired and confused in life, please stay at the shelter of our soul, then the vanity and noise will be left, only the silent dialogue as well as your heart remains...

当下，是相对过去而言，在物质逐渐丰裕之后，人们多追求的是如何在有限的空间内无限放大自己的情怀。卧室，承载了人生中一半的光阴，满足日常所需，也是心灵的栖息之地。一种闲适、潇洒的超脱情怀来自于张元干的词，用湖水荡漾、青山倒影作为此空间的意境再适合不过了。这空间用了大量的茶镜做为空间延伸的媒介，立体画中的折叠皮革双半圆不正是湖水荡漾的又一体现。

Now is compared with the past, while with the gradually-gained wealth, people are pursuing how to enlarge their emotions and feelings infinitely within the limited space. A bedroom houses a half time during one's life, where daily needs should be satisfied and the soul could be rested. Derived from the poem of Zhang Yuangan, there is a leisure and footloose flavor. Flowing waters and shades of mountains are proper to be the atmosphere in this space, where large amount of tawny mirrors are used to be the medium of the extension of space, while the two-double semicircles made of leather in the stereograph is another embodiment of the flowing water.

设计师旨在将现有被认知的中国元素设计进行进一步优化，结合国际时尚潮流，根据市场定位，融入现代观念，采用现代的表现手法，将东方特色巧妙地融入作品中，传达出一个有深厚历史文化底蕴的民族气质。

The designers aim at optimizing Chinese elements that have already been recognized further, combining with international fashion trends to blend in modern concept according to the market positioning, and integrating the oriental features ingeniously into the project through modern techniques, for what the designers want to convey is the national temperament with profound cultural deposits.

融会
INTEGRATION

中国历史悠久，文化底蕴深厚，其精髓在室内设计中应用广泛，中式传统室内设计，常常寄意于九宫格、水墨画、玉佩（玉石），现代中式设计将现代元素与传统元素结合在一起，以现代人的审美需求来打造富有传统韵味的事物，让传统艺术在这个空间得到合适的体现。

China has a long history with profound cultural deposits, the essence of which is widely used in interior designs. Chinese traditional interior design is often adopted with squared papers, ink paintings and jades, while modern Chinese design combines modern and traditional elements together to create something full of traditional flavor according to the aesthetic demands of modern people, embodying traditional arts appropriately in this space.

项目名称	中德英伦联邦B区24#楼04户型示范单位
设计公司	柏舍设计（柏舍励创专属机构）
项目地点	四川成都
项目面积	180 m²
主要材料	大理石、工艺玻璃、皮革、玉石等

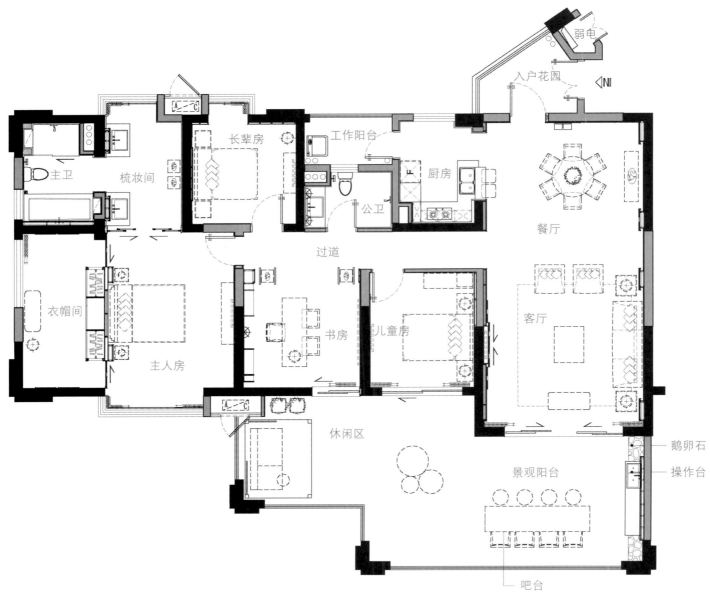

空间以橙色为主色调，增加空间的节奏感和现代时尚感，结合皮革家居表现了传统风格的典雅和华贵。客厅布置对称规整，主幅为手工水墨画的拼错，中间镶嵌九宫格，营造一种深邃的禅境。而在客厅的设计中，设计师将水墨画屏风置于空间一角，成为一处点睛之笔；餐厅以圆形餐桌寓天圆地方，别致的烛台吊灯，与整个空间主旨相呼应，温馨雅致。

This space is dominated by orange color which adds the rhythm and fashion, combined with leather furnishings, showing the elegance and luxury of the traditional style. The living room is symmetrical and ordered with a hand-made ink painting inlaid with squared papers in the middle, creating a profound Zen flavor. While in the design of the living room, the designer put the screen in ink painting on the corner of the space which became the highlight. The round table in the dinning room implies the round sky and the square earth, where the delicate candlesticks and droplights echo with the main theme, appearing warm and elegant.

书房延续了整体设计风格，设计师利用蒲扇的造型将翠竹以水墨画的形式呈现出来，黑白两色，素雅寂静，与对面墙面的山水倒影相映衬，伏案而作，鸟鸣啾啾，歌山诵水，打造一方自然幽静之地。本案的设计将两种风格元素融会贯通，中式元素，是国之精髓，在现代设计理念中，它给予我们更多思考生活，感悟生活的角度。

The study continues the overall design style, where the designer presents the bamboos in the ink painting through the modeling of a cattail leaf fan, the black and white colors of which are elegant and tranquil, foiled with the shades of the mountains on the wall, creating a natural and quiet place. The design in this case is the combination of two styles. Chinese elements are the quintessence of our country, which give us more angles to think about life among the modern design concept.

270度视角俯瞰都市繁华
270° VIEW OF URBAN PROSPERITY

项目名称 ｜ 中邦艾格美国际公寓样板房
设计公司 ｜ 上海柏仁装饰工程设计有限公司
设 计 师 ｜ 杜柏均
参与设计 ｜ 王稚云、季丹
项目地点 ｜ 上海
项目面积 ｜ 139 ㎡
主要材料 ｜ 钛金板、白色烤漆、金箔、硬包等
摄 影 师 ｜ 胡文杰

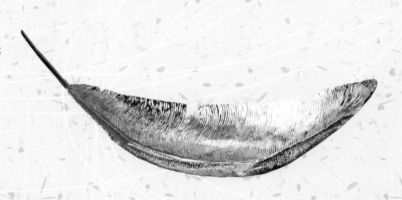

柏仁国际联合设计 总经理

    杜柏均，现上海柏仁装饰工程设计有限公司总经理，从业26年来一直秉持着将生活哲学融入个人设计的理念，其设计特点突出，注重彰显东方人的喜好，作品透露着丰富的人文主义精神。

    自2004年于上海创办上海柏仁装饰工程设计有限公司伊始，10年来杜柏均先后为证大集团、融侨集团、长甲集团、荣智健先生、君地投资集团、新华传媒集团、中润解放地产、华夏幸福基等合作了许多设计项目。如证大集团的海门九间堂样板房、证大大拇指广场商业综合体、证大大拇指广场酒店式公寓、证大大拇指广场办公楼、上海鹏欣集团白金府邸样板房等。

    其中，2015年APD亚太精英邀请赛获奖，2015年荣获金外滩奖最佳优秀售楼处奖，2014年中国室内设计年度杰出人物，2014年金堂奖获得优秀住宅样板房奖，2014年艾特奖获得优秀住宅样板房奖、最佳优秀陈设奖及TOP-50年度优秀设计师，2014年中国室内设计总评榜最佳住宅样板房奖、年度优秀设计师奖、2014及15年担任搜狐焦点居家设计师大赛评委等。此外，因其在室内设计上的成就和贡献，当选了由美国《室内设计》中文版、美国室内设计中文网及CIID中国建筑学会室内设计分会主办的"2014-2015中国室内设计年度封面人物"。

纽约、巴黎、伦敦、东京、上海……一路走来。放下行李箱的那一刻，唯有家的温暖，才真正让人感受到一丝人间的气息。那种充满肺腑的牵挂，倚靠在沙发上的无比舒适感，让一切都安静下来，只留下一个真正的自己。设计师为中邦艾格美国际公寓打造的样板房，没有浮夸和喧闹，与伦敦的精致、巴黎的时尚和上海的优雅相得益彰。如此缔造，让行走于世界的海归人士，萦绕在心头。

Along the way of New York, Paris, London, Tokyo and Shanghai, only when you put down your luggage, could you feel a little sense of real life. The caring from one's bottom of heart and the incomparable comfort when leaning on the sofa make everything calm down and there is only you remained. Designers create this show flat without any exaggeration and noise, which matches the delicacy of London, the fashion of Paris and the elegance of Shanghai perfectly, making the overseas returnees who have travelled along the world have something in mind.

有种视角 270度俯瞰都市繁华
There's an angle of view, 270°view of urban prosperity

对于一个拥有270度全景的室内格局来说，椭圆形的结构给予了设计师很大的挑战与机遇。为了尽可能的利用大平层的采光优势，全墙面的层柜设计，让阳光洒满了空间的长廊。没有阴暗角落的空间立刻变得自由通畅起来，原有的空间界定被魔术般的打破，立体的大视野让人有一种被环绕的氛围，无论从哪一个角度去感受，都无比舒适。

从玄关开始，入口处的管井就被巧妙的与收藏柜结合，自然形成天然的空间屏障，而这种屏障没有完全僵硬的封闭，透明的格子图腾设计，在若隐若现间，折射出空间的无穷。卧室则是百分百保留了弧形视野的落地窗，没有任何阻碍。智能科技的引入，让这种全景视角在夜晚也没有被遗忘。电动窗帘的采用，保障了主人可以毫无顾忌的俯瞰上海美丽的夜晚，丝毫不用担心偷窥。

For an interior layout with 270° full view, the elliptic structure gives designers a big challenge as well as an opportunity. In order to take advantage of the lighting in the big flat as much as possible, designers used the full-height cabinet to let sunshine break into the gallery in the space. The original dark corner suddenly becomes free and transparent, and the original boundary of space is broken magically, the large tridimensional field of which makes people feel as if they were surrounded to enjoy the scenery from any angle comfortably.

The tube well at the entrance of the porch is combined with the collection cabinet ingeniously which forms a natural partition, not being rigidly closed, while the transparent lattices refract the boundlessness of the space. The French window with arc view is kept without any obstruction. The introduction of intelligent technology makes the full view realized at night. The using of the electric curtain ensures the owner to overlook the beautiful night of Shanghai without peeping.

有种关爱 是自由的感受
There's a kind of caring with a feeling of freedom

作为驿站的居所，最大的目的是保障私人空间的无比自由。因此卫浴可以成为卧室开放空间的一部分，被深藏于闺阁的浴缸成为了"舞台"前的主角，她被安放在拥有最佳视野的落地窗前，陪伴主人度过每一个温馨的时刻。无论何时，都不会错过自然给予的美景。清新的日出，恬静的夕阳，又或是寂静的月光，都美美的存在着，被感知着，享受着。

As a residence, the biggest aim is to guarantee the freedom in the private space. Thus, the bathroom becomes a part of the open space in the bedroom, where the bathtub becomes the leading actor in front of the "stage" which is put before the French window with the best view, accompanying the owner to spend each warm moment. They could enjoy the natural scenery at any time. The fresh sunrise, the elegant sunset or even the silent moonlight are existed with their beauty.

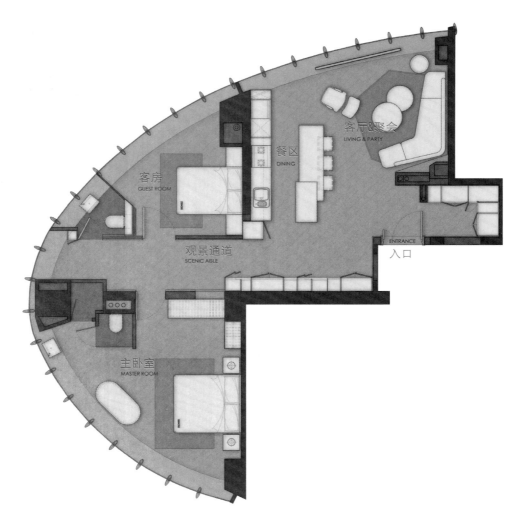

有种美 被称为精致
There's a kind of beauty called delicacy

  黄色的钛金板、白色的烤漆、鸵鸟皮的硬包……这些材质毫不吝啬的运用，让整个空间在金色、黑色和白色间转换，没有过多的浮夸，只有精致的塑造。客厅结合开放式的纽约大都会厨房，可以方便喜欢烹饪的主人开各种PARTY。呼朋唤友间，享受的不仅是快乐，也是一种生活方式的喜悦。

  极致的精致还体现在软装的搭配上，从金色的餐盘到镶嵌金箔的捷克水晶杯；从真皮沙发到皮草软垫；还有钢琴烤漆的茶几及吧椅，以及采用德国镶钻面料的窗帘和纱幔，又或是意大利顶级的卫浴龙头……这一切精彩的曼妙搭配，都体现着设计师的时尚与不凡的品位。

The yellow titanium board, the white stoving varnish and the hard decoration in ostrich skin...the using of these materials makes the whole space transform among gold, black and white colors with delicate shaping instead of any exaggeration. The living room is combined with an open kitchen in New York Metropolis style, which is convenient for the owner who loves cooking to hold parties where you could enjoy the happiness as well as joy of a kind of lifestyle.

The ultimate delicacy is reflected on the collocation of soft decoration from the gold plate to the crystal glass inlaid by gold foil; from the leather sofa to the fur cushions; as well as the tea table and barstools in piano baking varnish, the curtain and voile made by Germany lining inlaid by jewels, or the top Italian bathroom faucet...The wonderful collocations present the fashion and extraordinary taste of the designers.

毛明镜 Wally Mau

上海牧笛室内设计工程有限公司 创始人、高级合伙人

昌影 Alia Chang

上海牧笛室内设计工程有限公司 合伙人

　　Wally拥有敏锐的设计嗅觉和先进的管理理念思想，较高的空间整体把控能力，对细节臻于完美的性格促成了其高品质的审美眼光，是个完美主义追求者。在如今商业气氛浓厚的现代社会中，其始终努力寻求着设计与商业的平衡，设计的作品充满雅致之感，同时又不缺创意和惊喜。

　　Alia的作品尽显了她对建筑艺术的热爱，以及为客户设想的细心。在过去的10多年里，Alia的作品遍及国内一线知名开发商，令她对地产室内设计有了更深刻的体会与理解，并发展出自己敏锐的设计触觉。Alia不仅是一名经验丰富的设计师，更是一个优秀团队出色的领路人，在她的带领下，团队拥有较高的凝聚力，并源源不断的迸发出新的设计灵感。

感怀静时光
REMINISCENCE OF THE TRANQUIL TIME

跨界时尚
CROSS-BOUNDARY FASHION

感怀静时光
REMINISCENCE OF THE TRANQUIL TIME

路劲北郊庄园是路劲北郊城第五期，是路劲地产着重打造的标杆性项目。项目整体采用全石材立面，拥有超大面宽的法式联排、叠加。独特设计叠加、联排，均有天有地有花园，宽阔大气、尊贵经典，同时拥有高品质惬意生活。

Chateau du Nord is the fifth period developed as by RK Properties as a landmark. Stone facades are adopted in the whole project. The French townhouses and superimposed villas house large face widths which are unique with sky, ground and garden together, the generousness, nobility and classic of which make the inhabitants have qualified life leisurely.

项目名称	路劲北郊庄园
设计公司	上海牧笛室内设计有限公司
设 计 师	毛明镜、昌影
项目地点	上海
项目面积	216 ㎡
主要材料	石材、木饰面、硬包、皮革等

　　17世纪中期，明清易代的动乱给对外商贸带来契机，以此在欧洲诸多国家掀起了中国风的热潮。直到18世纪，东西方相遇，吸收华风的再创造，出现了当时非常流行的一种法式装饰艺术风格，Chinoiserie。它是中国文化在欧洲传播的典型代表，至今影响着今日的艺术装饰领域。本案设计沿袭着这一风格，在经典中寻求创新与突破，同时将实用与功能巧妙的隐藏在律动的线条之下。素雅的色彩、舒服的光调、精简的装饰，仿佛时光静止的刚刚好！

In the middle 17th century, the turmoil at the replacement from the Ming dynasty to the Qing dynasty brought opportunities for foreign trade, thus Chinese style began to be popular in many European countries. Until the 18th century, Chinese style met western style and then formed a French decorative artistic style called Chinoiserie, which was quite popular at that time. It was a typical example of Chinese culture that spread in Europe, which still influence the field of artistic decorations nowadays. Followed with this style, the designers sought innovations and breakthroughs from the classic, and hided the practical functions ingeniously under the rhythmed lines. The elegant colors, comfortable lights and concise decorations make people feel as if time stays quietly right here.

跨界时尚
CROSS-BOUNDARY FASHION

项目名称	景瑞宁波首南洋房
设计公司	上海牧笛室内设计有限公司
设 计 师	毛明镜、昌影
项目地点	浙江宁波
项目面积	300 m²
主要材料	石材、木饰面、金箔、手绘墙纸、硬包、夹丝玻璃、皮革等

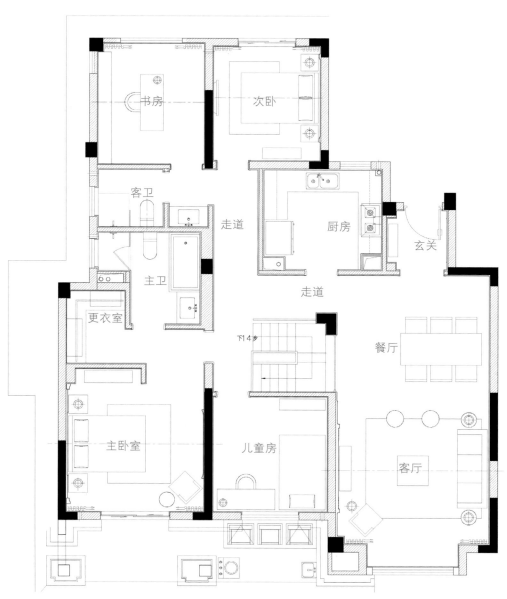

丰富、开放、创新的欧洲设计思想及其奢华尊贵的姿态一直备受追捧，新古典主义中保留了传统的历史痕迹及浑厚的文化底蕴，摒弃了纷繁复杂的装饰，将功能性与造型优美配搭，更符合了现代人的生活方式和生活态度。从某种程度上来讲，新古典主义是一种杂糅，是一种多元化思考方式下的产物，浪漫的怀旧气息与现代人生活需求融合，华贵典雅与时尚现代并行，这便是后现代时代个性化的美学观。本案设计师充分抓住了这些特性，完美演绎于整体设计中。

The rich, open and innovative European design idea with its luxurious and noble presence is always being chased after. Neo-classicism remains traditional historical traces and profound cultural deposits, abandons complex decorations and collocate the functions and graceful modelings together, which suits the lifestyles and living attitudes of modern people. In a way, neo-classicism is a mix as well as a product under diversified ways of thinking. The romantic and nostalgia sense is combined with the living demands of modern people, while the luxury and elegance stay together with modern fashion, which is the individual aesthetics in the post-modern age. The designers in this case seized these features and interpreted them into the whole design perfectly.

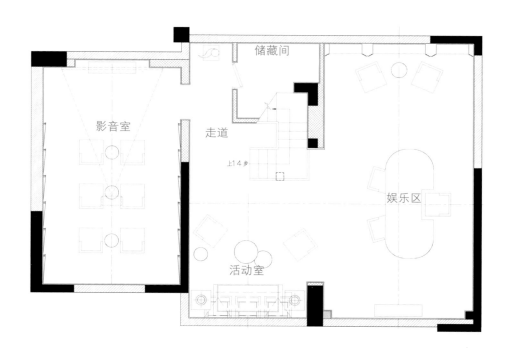

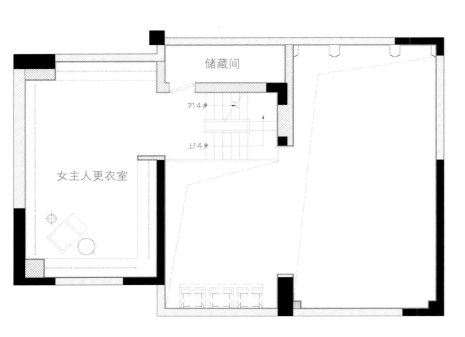

储藏间

下14步
上14步

女主人更衣室

王兵 / BING WANG

上海无相室内设计工程有限公司 创办人、设计总监

王兵，毕业于上海同济大学建筑系，从事室内设计25年。不受风格与形式束缚，秉持"传承中创新"的设计理念，深度解读现代与古典美学并以独到手法演绎，从功能出发创造出具有生命力的场景。认为空间环境既具有使用价值，同时也应反映相应的历史文脉、建筑风格、环境气氛等精神因素，室内设计的终极目的是"创造满足人们物质和精神生活需要的室内环境"。

与美国JWDA合作的苏浙汇1933店曾荣获美国Boutique Design Awards室内设计大奖金奖，连续三届获上海装饰装修行业协会室内设计大赛金奖，亚太室内设计大赛荣誉奖等国内外10多个设计奖项，苏浙汇虹桥店曾获香港杂志评为"年度最佳餐厅"。

来自北欧的清风
REFRESHING BREEZE FROM NORTHERN EUROPE

意式柔情 多彩之歌
ITALIAN TENDERNESS WITH COLORFUL SONG OF PRAISE

来自北欧的清风
REFRESHING BREEZE FROM NORTHERN EUROPE

项目名称	苏州昆山和风雅颂样板房
设计公司	上海无相室内设计工程有限公司
设 计 师	王兵、徐洁芳
项目地点	江苏苏州
项目面积	300 ㎡
主要材料	白色乳胶漆、橡木原木、白色玻化砖、壁纸等
摄 影 师	张静

亚太豪宅大赏 III · 风靡大陆 | 127

北欧清风拂面而来
Refreshing breeze from Northern Europe

北欧风格重在体现出简约、舒适之感,整个空间摈弃过多装饰色调,干净清雅,以纯白色、原木色及温和的灰蓝色、浅褐色为主,白墙、原木、创意灯具,恰到好处的留白与尺度营造出恬静宜人的生活场景。真正的品质往往隐藏在看不见的细节里,这里的沙发、茶几、餐桌、床等家具多数为量身定制,所用的实木原料在仓库里养了近一年,自然烘干以后做出来的家具极富质感,充分展现出木材固有的温情与优美。

Northern European style emphasizes on embodying the conciseness and comfort without complex decorative colors, yet dominated by pure white, wood, gray blue, light brown and other clean and gentle colors. The white walls, raw woods and creative lamps create a tranquil and comfortable living scene. True qualities are always hided in some invisible details. The sofas, tea tables, dinning table, beds and other furniture are mostly custom designed, the solid wood of which is natural and qualified, revealing the tender feelings and gracefulness of wood. concept of "inheritance and innovation" deeply.

设计师精心挑选每一件器皿、花艺、装饰品,营造出原汁原味的北欧风情。憨态可掬的小木鸟或在圆桌上成群蹲守,或在树枝上独自停驻,带来一幕幕生动有趣的画面;墙上的驯鹿头像、维京题材油画,排列整齐的圆木,则将人不由自主地带入北欧原始、质朴的生活情境中,行走室内,仿佛时时有北欧的清风拂面而来。

The designers select deliberately on each utensil, floriculture and decoration, creating the original taste and flavor of Northern Europe. The small charmingly naive wooden birds bring lively and interesting scenes; the head portrait of reindeer and Viking paintings on the wall and the ordered logs bring people into the primitive and pristine life of Northern Europe involuntarily as if there were refreshing breeze from Northern Europe constantly when walking indoors.

东方意境与北欧情调
Oriental flavor with Northern European style

"好的建筑是从土里生长出来的。"现代建筑领域里，美国建筑大师弗兰克·劳埃德·赖特提出的"有机建筑"理论将建筑视为树木，要扎根土地，与环境一体，有自己的生命力。这种贴近自然的思考将建筑和环境、和风土文化紧密联系在一起，以科学、合理的功能性探索来实现，也遥遥呼应了东方自老子以来所倡导的自然哲学。建筑要契合环境，好的室内空间同样要与建筑浑然一体，设计师正是基于这一原则，从使用功能、生活习惯等人居角度出发，在现代建筑背景中给出舒适、好用又兼具美感的生活方案。项目建筑为现代中式风格，白墙院落、双开钉门、砖砌照壁等元素从传统中式建筑简化而来，住宅内部构造遵从现代生活所需，布局简洁明了。在设计师王兵看来："越简单的空间包容性越强。北欧的简约、自律精神，其实在某种程度上跟简练的中式风格有些相似。"王兵很擅长从不同中找出共性，这种共性可以是近似的，也可以是互补的，其中精妙难言，而成功混搭多元要素的关键就在于此。借助温馨原木、素净的棉麻织物、玻璃器皿、花艺等元素，室内充盈自然清新的北欧风情；地下活动室则设有茶室，引入天井庭院里的松、竹、枯山水砂石，东方意境与北欧简约交融共生，创造出一种全新的素雅意境。

"A good building comes out of earth." In the field of modern architecture, Frank Lloyd Wright, the master architect from America, suggested the theory of "organic architecture" that architectures should be regarded as trees which need to be rooted in earth and stay together with the environment with its own vitality. This thought blends architecture closely with the environment and endemic culture which is realized through the scientific and reasonable functional exploration, echoing with the natural philosophy advocated by Laozi from the east. An architecture should be connected with the environment, while a good interior space should be integrated with the architecture as well, based on which the designers brought out a comfortable and beautiful plan within modern background from the aspect of functions, living habits and other ways. The building is adopted with modern Chinese style, the white walls, double doors, screen walls of which are simplified from traditional Chinese architectures, while the interior structures are concise and clear according to the demands of modern life. In Wang Bing's opinion, "The more concise the space is, the more inclusive of it. The concise and self-disciplined spirit of Northern Europe is similar with the simple Chinese style to some degree." He is good at finding the generality from the differences, which is either similar or complementary on mixing various elements. Through raw wood, clean ramie cotton fabric, glass wares and floriculture, the interior space is filled with natural and fresh Northern European flavor. There is a tea room in the activity space underground, where the pines, bamboos, dry landscape and dinas bring the oriental flavor and Northern European simplicity together, creating a new elegant atmosphere.

和自然无缝衔接
Connected perfectly with nature

"设计一个家,最重要的是舒适、好用、安全,每样东西的功能性琢磨透了,做到最合理,自然而然就会好看。"王兵在设计这套样板房时,尝试去了解北欧归来的学子们会有的一些西方生活习惯,在功能设置时将这些习惯以创新的手法加入,让空间既传承北欧特色,又符合中国式生活习惯。

其中亮点之一是根据北欧当地的生活习惯,将浴缸设计在卧室床边,泡澡时可惬意享受庭院绿色美景。另一个亮点是地下层的茶室与Spa区。茶室因中国人生活习惯而设计,Spa区则是因为气候环境的因素,桑拿、壁炉都是北欧人生活里不可缺少的部分,设计师设计的时候特别考虑到了景观的加入。Spa池临窗而设,正对通达地下层的天井庭院,并且特意向下拉长窗框,以窗为一侧池壁,巧妙的框景效果让室内外充满宁静禅意。

"When designing a home, comfort, convenience and safety are the most important. If the functions of each are reached to the most reasonable, it will be wonderful naturally." During the design of this case, Wang Bing tried to understand the western living habits of these returned students who had studied in Northern Europe and added these habits into the space via innovate techniques, inheriting the features of Northern Europe as well as conforming to the living habits of Chinese style.

One of the highlights is designing a bathtub beside the bed according to the living habits of Northern Europe, where you could enjoy the beautiful sceneries when bathing. Another highlight is the tearoom as well as Spa area underground. The tea room is designed as the living habits of Chinese people, while the Spa area is considered according to the environmental factors since sauna and fireplace are the necessary parts in Nordic lives. The Spa pool is settled by the window towards the patio underground with a stretched window frame, making both the outdoor and indoor space filled with the tranquil Zen flavor.

当北欧邂逅中式,诗意的栖居
Idyllic living between Northern European style and Chinese style

本案从属于政府的留学归国人才奖励项目,其中包括一批从北欧归来的留学人士,因此空间从一开始就定为简约、自然、实用的北欧风格。室内地面二层,地下一层,设计师王兵除了在色调、材质、装饰方面原味呈现出现代北欧风情之外,还从居住者的生活习惯出发,对动线安排、空间互动方面进行了独特规划,深度展现了他"传承与创新"的设计理念。

This case is attached to the government project for the returned talents, including those studying in Northern Europe, thus it is settle as the simple, natural and practical Northern European style from the beginning. There are two floors and one floor underground. The designer Wang Bing presented modern Northern European style on the colors, materials and decorations and planned the generatrix and interactions uniquely from the living habits of the inhabitants, revealing his design concept of "inheritance and innovation" deeply.

意式柔情 多彩之歌
ITALIAN TENDERNESS WITH COLORFUL SONG OF PRAISE

本案设计师综合考量外部环境、建筑风格、业主定位等多方面因素，给出的设计不仅传承了意大利文化丰富的艺术底蕴，也融入了开放、创新的设计思考，因地制宜，东西交融，在传统中推陈出新。

Considering the exterior environment, architectural style, orientation of the owner and other factors, designers in this case provided a design which both inherited the rich artistic deposit of Italian culture, and combined the open and innovative thought, getting rid of the stale and bringing forth the fresh through the integration of eastern and western elements.

项目名称	西安卢卡小镇欧式别墅样板房	项目地点	陕西西安
设计公司	上海无相室内设计工程有限公司	项目面积	310 ㎡
主创设计	王兵	主要材料	石材、白色浑水木饰面、壁纸等
参与设计	谢萍、李倩	摄影师	张静

本案位于陕西新设立的西咸新区，所在楼盘位居泾河北岸，崇文塔畔，地处西安的南北主轴线上，既保有自然田园风光，又贴近强劲都市脉动。从地理环境深入内蕴人文，设计师王兵将温情闲适又浪漫优雅的意大利情调融入其中，开放多元的设计思考在传承中创新，通透明朗的空间设置、化繁为简的造型线条、缤纷柔和的多重色彩，让人如同置身于充盈多彩柔情的波托菲诺海滨，进驻令身心宁静的心灵港湾。

This case is located at the newly built Xixian New Area of Shaanxi province, all the premises of which are set on the north bank of Jing river beside the Chongwen Tower on the north-south principal axis of Xi'an, with natural idyllic scenery as well as urban life. The designer Wang Bing blended the gentle and leisure yet romantic and elegant Italian flavor into the house, where the open and diversified design thoughts were innovated within the inheritance, the transparent and bright settlement of space, the concise modeling lines and the gentle colors of which made people feel as if they were living in the colorful and tender shore of Portofino, enjoying the tranquil harbor in mind.

东方托斯卡纳

Tuscan style in the east

这套联排别墅隶属西咸新区泾河新城的"卢卡小镇"，整体引入意大利托斯卡纳建筑风格，在西安"后花园"打造融汇东西方文化的宜居现代田园。托斯卡纳的金色阳光、红色土壤、苍翠葡萄园孕育出引领欧洲文明的文艺复兴浪潮，它是田园的、闲适的、也是优雅的、浪漫的，是令人憧憬的心灵栖息地。这种内蕴的人文精神可以超脱时间、跨越空间，在中国内陆的千年古都同样能够引发情感共鸣，它是本案建筑骨架的依凭，也是其室内设计的出发点。

This townhouse is located at the "Lucca Town" of Jinghe New City in the Xixian New Area, introduce by Italian Tuscan architectural style, creating a modern livable place with the combination of eastern and western culture at the "backyard garden" of Xi'an. The gold sunshine, red earth and verdant vineyard of Tuscany breed the trend of the Renaissance in European civilization, which is idyllic, leisure, graceful and romantic, even the longing home of soul. This intrinsic humanistic spirit is beyond any time, which could be resonated in an ancient capital for thousands of years in the mainland of China. This is the basis of the architectural framework as well as the starting point of the interior design in this case.reindeer and Viking paintings on the wall and the ordered logs bring people into the primitive and pristine life of Northern Europe involuntarily as if there were refreshing breeze from Northern Europe constantly when walking indoors.

意式传统新演绎
New interpretation of traditional Italian style

空间布局是整个设计的起点。这套联排别墅地上三层，地下一层，客厅、餐厅、厨房等公共活动区设于一层，起居娱乐区在地下层，二层设两间客卧套房，三层即为主卧套房，附带观景露台。310平方米的居住空间功能设置齐备，为了呈现更为年轻化的配置，使其生活方式和形态也都更趋向于现代感，空间感受上格外注重通透、自然、舒适、方便。为此，一层将门厅、两层楼高的挑空客厅、餐厅与西厨房打通，融为一体，只是通过对称的门洞作为空间分割的界定。地下室活动空间内，起居室与酒吧合二为一，形成一个更欢乐的娱乐空间，而相对独立又有关联的健身房、瑜伽室及水疗区又为时尚健康的生活增加了一个中心亮点。

The whole design is started from the space layout. This is a townhouse with three floors above ground and one floor underground. The living room, dinning room, kitchen and other public activity areas are settled on the first floor; the area for daily life and entertainment is settled on the basement; the second floor houses two guest suites while the third floor houses the master suite with a viewing balcony. This living space with 310 square meters is equipped with many functions. In order to present a younger configuration and make the lifestyle more modern, it emphasizes especially on the transparency, nature, comfort and convenience. Therefore, the hall, high-ceilinged living room with two floors' height, dinning room and western kitchen on the first floor are all opened up, while symmetrical door openings are remained to be the demarcation of the spaces. The sitting room and bar are integrated in the basement, forming a happy

entertainment space, while the relatively separate yet related gym, yoga room and spa area add a highlight into the fashionable and healthy life.

我们可以看到室内的空间形态是遵循章法的，新古典主义风格的墙地面细部处理也中规中距，一方面保留了材质、色彩的大气风范，仍然可以很强烈地感受传统的历史痕迹与浑厚的文化底蕴，同时又摒弃了过于复杂的肌理和装饰，简化了线条。素净石材与白色浑水木饰面给予明净爽朗的背景基调，镜面装饰的运用进一步拓展空间感，从挑空客厅、餐厅到地下起居室、楼上卧室都可见不同形态的镜面镶嵌，特意选用的锈镜赋予空间岁月沉淀的年代感。

We could see that the interior spatial form is followed with principles, where the handling of the walls and ground with neo-classical style are upright, remaining the generous manner of the materials and colors to make people feel the traditional historical traces and profound cultural deposit as well as abandoning the complicated texture and decorations and simplifying the lines. The plain stone and white wooden veneer give a clean and clear background, while the using of mirror adds the sense of space further. You could see mirrors in different forms from the high-ceilinged living room and dinning room to the sitting room underground and even the bedroom upstairs, while the selected rust mirror endows the sense of time.

多彩波托菲诺
The colorful Portofino

缤纷鲜活的多重色彩是意大利度假胜地波托菲诺给人留下的第一印象。色彩斑斓的错落屋舍背靠绿意盎然的山丘，面朝碧波浩淼的大海，绿树、红花、蓝天、碧海、金色阳光，空气中弥散着热情自由的味道，充盈着明朗温馨的气息。这丰饶微妙的自然色彩成为本案空间配色的灵感之源。以白色为基底，奶油色、浅灰、大地棕妆点在墙面、地毯、窗幔上，绿与黄、金与蓝则碰撞跳跃于定制的沙发、座椅、靠枕上，嫩黄的花束、金黄的烛台与艺术陈设，如同灿烂阳光，点亮此间的都市田园生活。

The first impression of Portofino, the famous Italian resort, is the lively colors. The houses with diversified colors against the green hills are facing to the sea, where green trees, red flowers, blue skies, seas and gold sunshine pervade an enthusiastic and free flavor, filled with the bright and warm sense. These natural colors become the source of inspiration in this case which is based on white and embellished by cream color, light gray and brown on the walls, carpets and curtains, while green and yellow, gold and blue are jumping on the custom designed sofa, chairs and cushions, the bright yellow flowers, gold candlestick and artistic furnishings lighten this idyllic life in the city like sunshine.

壹城壹墅
ONE CITY ONE VILLA

项目名称	珠江壹城A5区G10-02单元别墅样板房
设计公司	广州共生形态创意集团 www.cocopro.cn
设计总监	彭征
设计团队	彭征、谢泽坤、吴嘉
项目地点	广东广州
项目面积	410 ㎡
主要材料	大理石、玫瑰金、玻璃、皮革、木饰面、墙纸等

广州共生形态创意集团 董事、设计总监

  共生形态创始人、设计总监，广州美术学院设计艺术学硕士，曾任教于华南理工大学设计学院，现为广州美术学院建筑艺术设计学院客座讲师、实践导师，中国房地产协会商业地产专委会商业地产研究员。

  彭征先生关注城市化进程中的当代设计，主张空间设计的跨界思维，从事建筑、室内、景观等多领域的设计实践，设计作品具有较强的建筑感和现代简约的风格。设计作品曾获德国红点设计大奖、美国室内设计杂志Best of Year Awards年度最佳大奖、意大利 A'DESIGN AWARD银奖铜奖、香港亚太室内设计大奖、金堂奖、现代装饰国际传媒奖、艾特奖等国际和国内设计大奖等。

　　珠江壹城是珠江地产投资380亿元进行连片开发的超级城市综合体,整个项目总规划占地面积约1.9万亩,本项目室内设计及软装陈设由广州共生形态设计倾力打造。

　　别墅根据楼层和功能分为三层:负一层娱乐空间,包含红酒区、健身房和豪华SPA区;一楼为起居空间,包含7米中空的客厅,开放式厨房、餐厅和茶室三位一体的就餐区以及老人房和户外花园;二楼设主人套房和女儿套房,南北对流,户户有景。

Pearl River One City, a super-city complex developed by Pearl River Real Estate with an investment of 38 billion yuan, covers an area of 19 thousand mu in the plan of the whole project, the interior design and furnishings of which are created by C&C Design.

According to the floors and functions, this villa is divided into three layers: the basement is an entertainment space, including a wine area, a gym and a luxurious SPA area; the first floor is a space for daily life, including a high-ceilinged living room with 7 meters high, a triune dinning area with an open kitchen, a dinning room and a tea room, as well as an elder's room and an outdoor garden; while the second floor includes a master suite and a daughter's suite. There is a north-south convection in the house where scenery could be seen in every room.

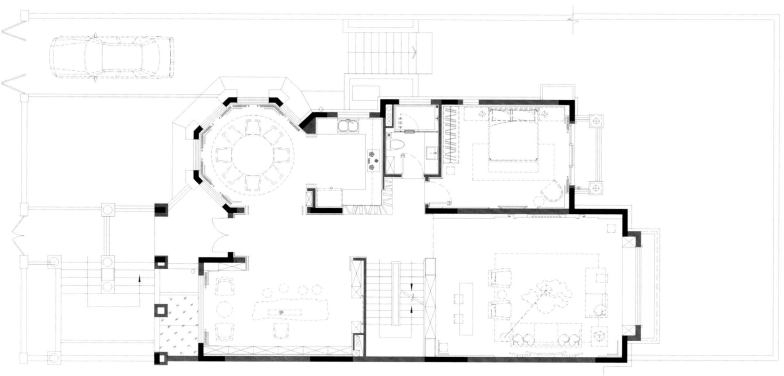

在此空间塑造中，设计师以现代人的生活方式融入东方气质美学营造意境，并尝试用跳跃的色彩和时尚审美融入具有人文底蕴的简约中式风格之中，生活细节的精到把握，又于共生中见生活美意。

In the space creating, the designer blended oriental aesthetics through the lifestyles of modern people to create the artistic conception, and tried to apply jumping colors and fashion into the simple Chinese style with cultural deposit, reappearing the beauty of life via the ingenious handling with the details.

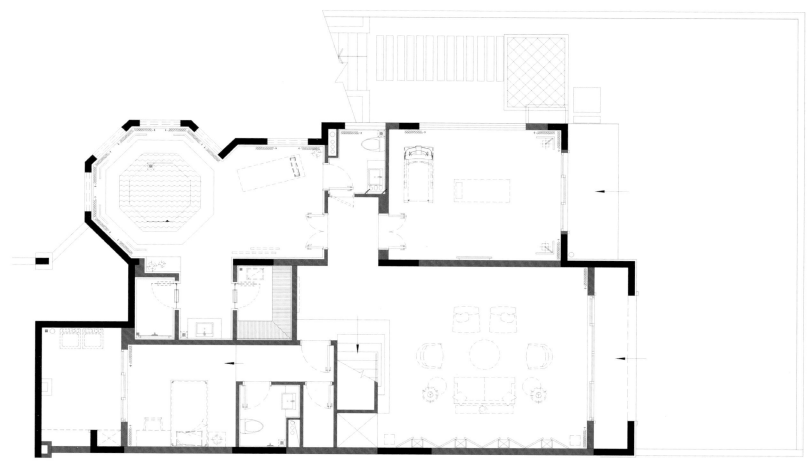

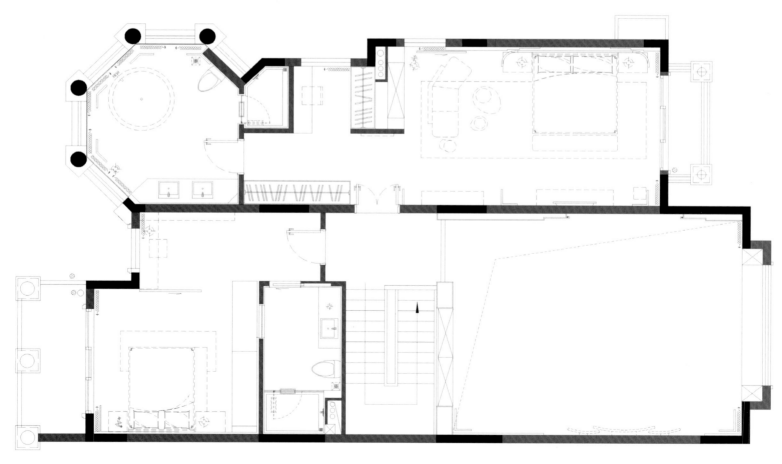

壹挚设计集团 C&C Design Group

C&C壹挚设计集团（以下简称C&C），是由一群具有国际先进理念的精英人才组成的综合性文化设计机构。C&C始创于2003年，旗下包括广州市壹挚室内设计有限公司，卡络思琪装饰设计有限公司，玲珑堂家品，香港利睿财富管理有限公司等数个跨领域的专业服务品牌。

摩登雅痞
MODERN YUPPIE

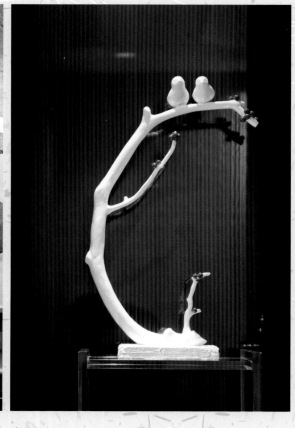

摩登东方
THE MODERN ORIENT

摩登雅痞
MODERN YUPPIE

项目名称	华标品雅城一期别墅B型
室内设计	C&C壹挚设计
软装设计	C&C卡络思琪
主设计师	陈嘉君、邓丽司、贺岚
项目地点	广东广州
项目面积	365 m²
摄 影 师	谢艺彬

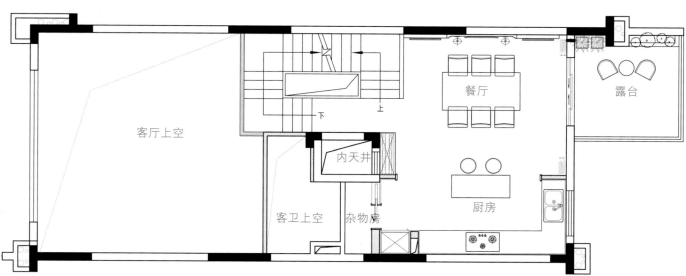

　　本案灵感源于优雅至极却又悠然随意的贵族气质，对现实生活墨守成规的大胆突破，把平行不相干的元素巧妙糅合起来突破陈规的枷锁，形成耳目一新的感官享受。这是一个三层高的别墅，空间十分宽敞明亮，有足够的生活空间之余可以格外满足地享受私人生活。一层是挑高的客厅，别墅总体均采用棕色系的设计，客厅则自然沿袭同样的色调，中式的格纹和美式生活里偏好的皮质家具甚至是金属装饰相碰撞，简洁大气的设计不会使感觉游走至不知情趣的油腔滑调，再加上加长白色沙发的调和，通身显露出十足的大户人家气派。

The idea of this case is derived from the graceful yet leisure aristocratic temperament. Designers broke the rules in real life bravely and combined the irrelevant elements together ingeniously, forming an entirely new feeling. This is a spacious and bright villa with three floors, where one could enjoy private life contentedly beside daily activities. The first floor houses a high-ceilinged living room with brown tone which is used throughout the villa, where Chinese plaids collide with leather furniture or even metal decorations preferred in American life. The concise and generous design would not let the feeling go away, added with the compromise of the extended white sofa, revealing the generousness of a rich and influential family everywhere.

　　夹层是开放式的厨房和餐厅，以经典耐看的米灰色作为背景色，一抹高调冷艳的橘色透露出点到为止的精致，让贵族感溢满整个空间，有古典成熟的格调又有前卫端庄的气味，而更大胆的手法是在这份古典端庄中加入了偏工业风的金属装饰却丝毫不违和。冷暖色调的跳跃搭配丰富整体设计的层次感，搭配看似混乱无章实质蕴含奥秘的图案，加以动物元素点缀，让每个家品都能成为无可挑剔、不可或缺的艺术品。

The interlayer houses an open kitchen and a dinning room, dominated by the classical griege, embellished by a little cold orange color, showing the delicacy and making the nobility filled in the whole space accompany with the classical and mature style as well as the trendy and elegant flavor. It appears quite natural when metal decorations in industrial style is added into this kind of elegant flavor. The combination of both cold and warm tones enriches the layers in the overall space, collocated with chaotic yet mystical patterns as well as animal elements, making each furnishing at home become a perfect and indispensable artwork.

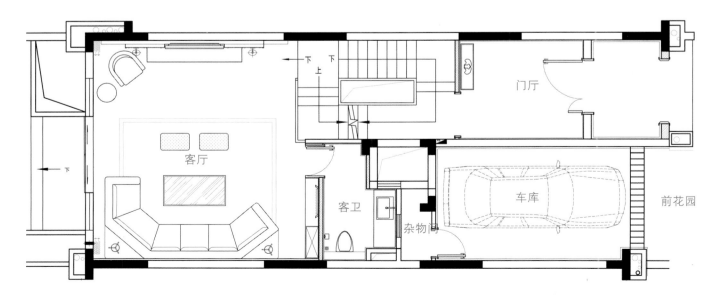

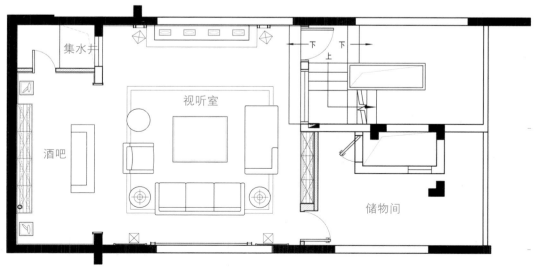

除此之外还设影音室，位于负一层，在大理石制造的十分豪华的视觉背面，还有很贴心的设计，安排了长沙发、单人椅子以及躺卧的椅子，可选择以最舒适的姿态来欣赏视听以及天花板的满天星河。

In addition, there is a media room at the basement, where designers arrange a long sofa, a single chair and a lounge at the back of the luxurious visual background made of marble, then one could choose the most comfortable posture to enjoy the audio-visual feast and the sky with stars at the ceiling.

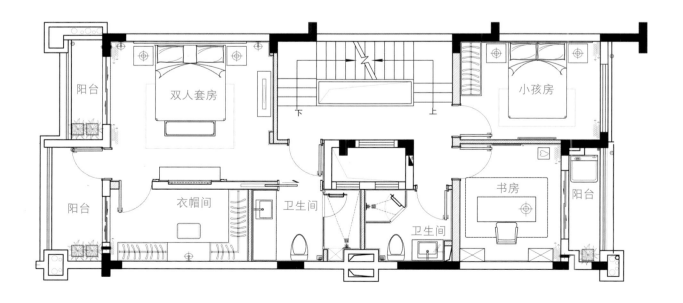

　　位于三层全层的是精心设计的主人房，设有衣帽间、书房、独立主卫以及超大阳台，混搭着的抽象艺术表现在这个空间里面显得那样的和谐，满足所有可令生活更美好的需要的同时增添一点有趣的装扮。爽朗大气的线条勾勒出干练的贵族气息。美式的紫铜、铆钉、豪气奢华的水晶大吊灯和富有质感的软麻布搭配出丰富的层次感，传递专属美式的张扬、豪放、向往自由的风格。设计师根据使用者的喜好，加入马术等动物元素，复古却又生动，一静一动，交相辉映。

The third floor houses the master bedroom with a cloakroom, a study, a private bathroom and a supper balcony which is designed deliberately, where the mixed abstract arts appear very harmony to meet the demands that could make life better as well as adding a little funny decorations. Generous lines depict clean and clear noble sense. The collocation of American red copper, rivets, luxurious chandelier and textured soft linen brings rich layers and conveys the bold, unconstrained and free style. According to the preferences of the inhabitants, designers add horse and other animal elements, which are vintage yet vivid.

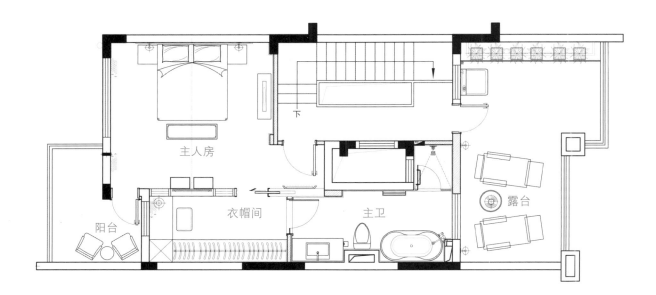

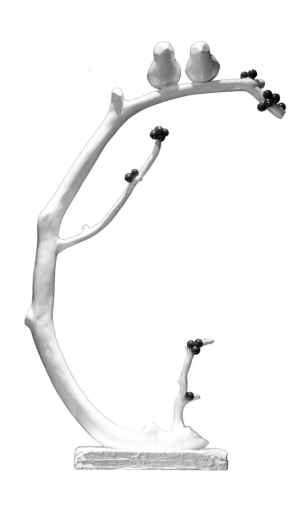

摩登东方
THE MODERN ORIENT

被悠久的文化浸润成长的人都有点古典情怀，这慢慢演变成一种生活需要，把这种需要展现在家居上，足可以在最舒适的场所和自己的生活有强烈的共鸣，增加内心的满足感。这是一个三层的别墅，一二层是起居室和餐厅，三层是主人家的卧室和主要活动空间。本案追求高雅简洁却又低调的淡雅气质，设计师运用现代东方主义设计，以耐人寻味的卡其色作为主背景，一改往日繁琐冗杂的修饰手法，用沉稳的笔触勾勒出温婉典雅气质，糅合空间的色调，一切仿佛浑然天成。相近色系完美的跳跃搭配使得空间有着优雅的变化，随处都展示着主人超脱的审美观和悠闲淡然的气质。

项目名称	华标品雅城一期别墅C型
室内设计	C&C壹挚设计
软装设计	C&C-卡络思琪
主设计师	陈嘉君、邓丽司、文斌华
项目地点	广东广州
项目面积	480 m²
摄影师	谢艺彬

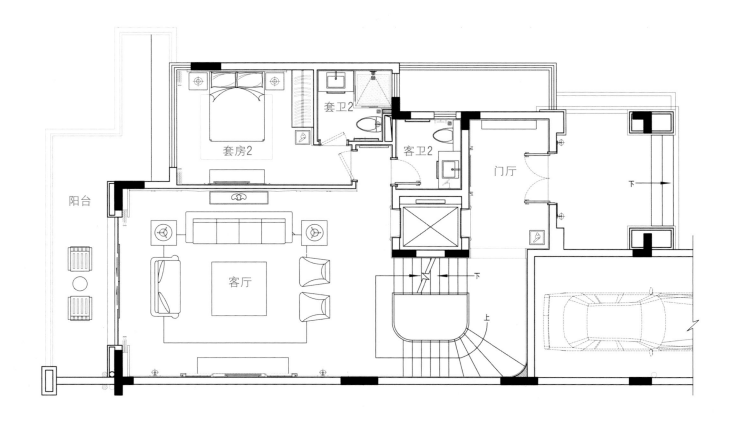

People who grow up with a long—standing culture have classical feelings more or less, which has been evolved into a demand in life gradually, which could produce a strong resonance with one's life when being presented on the furnishings in the most comfortable place, adding the satisfaction in mind. This is a villa with three floors, the first floor and second floor house a living room and a dinning room, while the third floor houses the master bedroom and the main activity space. It pursues for the elegant, concise and fresh temperament which is low-profile as well. Designers used the modern oriental design and a main background of khaki color, and outlined the gentle and elegant temperament through steady lines instead of the conventional complex techniques on decoration in order to mix the tones in the space naturally. The perfect match of similar color tones makes the space have graceful changes, presenting the owner's outstanding aesthetics and leisure temperament everywhere.

　　首层的客厅主要运用灰色以及卡其色，以悠久的东方文化作为基底，加以极具国味的泼墨和挂画，令其拥有持久感染力的气质。因而客厅在几种颜色和材质的合并调和中，表达出生活的一份优雅，反而有古典的东方情怀。餐厅位于夹层，刚好可以借来视野俯视客厅全景。而卧室的设计以简洁大气为主，沿袭屋子的棕色主色调，丝绸可让整个房间看起来明亮和有格调。

The living room at the first floor is dominated by gray and khaki, based on the oriental culture with a long history, and added with the inking paintings with strong Chinese flavor, giving the charming temperament for a long time. Thus, the living room expresses an elegance in life through the combination of several colors and materials with a little classical oriental feeling. The dinning room is located on the interlayer, which could be used to overlook the full view of the living room. While the design of the bedroom is mainly concise and generous with the brown tone followed with the whole space, the silk of which makes the room appear bright and stylish.

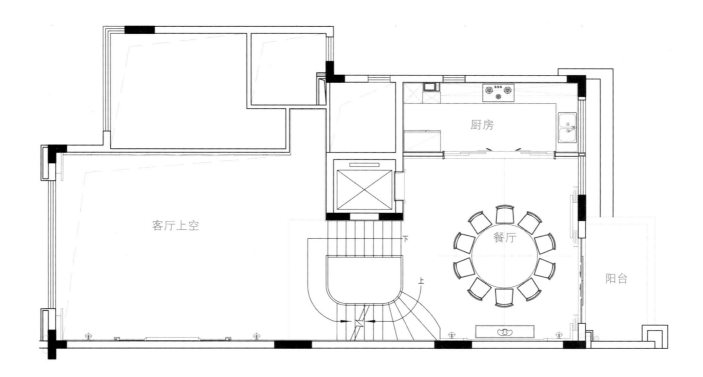

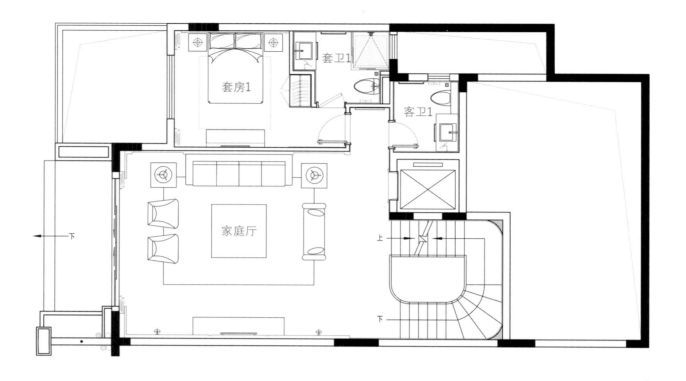
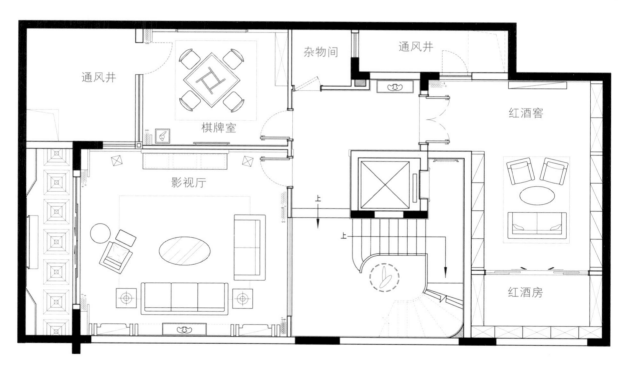
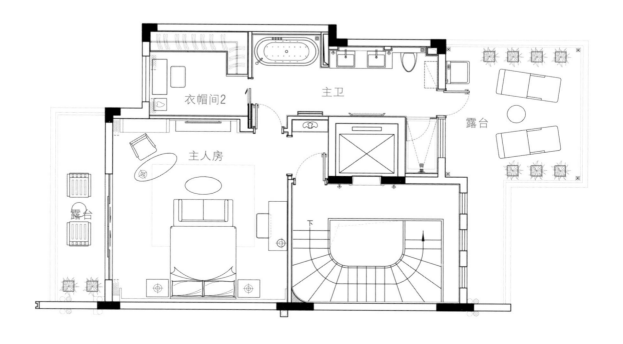

三层的主人房空间较大，设计了多功能使用的空间，有书桌为主导的小型工作角落和用于休闲的小型起居空间，在这里可实现多种活动，方便且舒适。这里设两个露台，从卧室走出去的小型阳台，以及从主卫走出去的豪华露台。主卫的设计已经脱离传统，出于生活需要的概念变成了一个享受生活的空间，加之露台的配合，全部设计如同行云流水，细微处流淌着极富变化的层次对比，却又过渡得悄无声息，透露着现代东方的沉稳与自信。

The master bedroom at the third floor is large enough to house some multifunctional spaces, such as a small working place leaded by a desk, and a small leisure space for daily life, where many activities could be realized with convenience and comfort. There are two terraces, one is the small balcony outside the master room, the other is the luxurious terrace outside the master bathroom. The design of the master bathroom is out of traditions to be an enjoyable space according to the demands of life, collocated by the terrace, presenting changeable layers smoothly and naturally with an invisible transition, revealing the steadiness and confidence of the modern orient.

圆舞曲
A WALTZ

设计公司	ACE谢辉室内定制设计服务机构
设 计 师	谢辉、王琦琅
设计团队	李蔓君、吴建波
项目地点	四川成都
主要材料	大理石、窗帘、木作、镜子、壁纸等
摄 影 师	李恒

谢辉 / ALYSSA

谢辉室内定制设计服务机构 设计总监

　　2010年,在经历了十年的设计生涯后成立了自己的工作室。最初是一个人的独行女侠,慢慢有了帮手。一名、二名,到现在的十人。开始做设计时的想法很简单,现在告诉自己设计是我一生的事业。我认为设计一定是有理想的,我的理想就是在未来,ACE能够成为行业的标签,代表着高品质与高品位。我的团队可以不大,但一定要是行业最精良的团队。为此,不断积累并专注于每个当下的项目我一直在努力。

曾就读于意大利米兰理工大学国际奢华酒店设计与设计管理大师班
成都市建筑装饰协会理事
CIID成都木兰会会员
2014年成都第十五届建筑装饰空间艺术设计大赛银杏金鸟奖金奖
2014年成都第十五届建筑装饰空间艺术设计大赛样板房空间提名奖
2014年成都第十五届建筑装饰空间艺术设计大赛工程别墅空间提名奖
2014年金堂奖年度优秀住宅公寓设计
2014年作品入选《2014中国室内设计年鉴》
2014年第五届筑巢奖别墅空间类优秀创意奖
2014年第五届筑巢奖公寓类优秀奖
2014年成都空间创意设计展最佳设计奖
2013年四川创意设计年度总评榜"十佳装饰设计单位"
2013年年度国际空间设计大奖艾特奖最佳别墅豪宅设计入围奖
2013年年度国际空间设计大奖艾特奖最佳文化空间设计入围奖

这里是一对夫妇为自己的三个孩子打造的一处梦想乐园，姐姐，妹妹还有不到一岁的小弟。本案动线安排均围绕三个孩子的生活展开，一层的家庭书区，音乐舞蹈室，二层的起居室与玩具室，三层的卧室及作业区。每个区域相对独立，基本围绕孩子们的生活学习而设，家是每个孩子梦想和爱的起点，每个孩子都是唯一的天使，在家庭的怀抱和浓浓的爱中滋养长大是每个父母的愿望。我们用孩子的视角，引用大自然的语言，用丛林、海洋的宽广介入室内，家的每个角落都流露对自然和美好的崇尚。一家人在书吧中享受读书时光，孩子们可以在玩具室接待自己的小伙伴，当然还可以从玩具室直接滑到一层的书吧！喜欢跳舞的妹妹可以在舞蹈室翩翩起舞，或者在三楼的作业室安静的完成作业，也可自由在墙上涂鸦。让每个孩子在家里都能找到乐趣和自由的存在，让童年地美好留在记忆的最深处，让亲情与爱包容每个家庭成员的成长，让孩子长大后不忘儿时的美好时光！

This is a dream land built by a couple for their children, two sisters and a baby boy. The generatrix in this case is surrounded with the lives of the children, such as the family reading area and music & dance room on the first floor, the sitting room and play room on the second, and the bedrooms and homework area on the third. Each area is relatively separate according to the lives and studies of the children. Home is the starting point of the dream and love in each child's mind, while each child is the only angel, thus letting them grow up in the arms of family with deep love is the wish of each parent. Designers used the children's angle of view, introduced the language of nature, and blended the broadness of forests and seas into the room, pervading the advocation towards nature and beauty through every corner. The family could enjoy their time reading in the book bar, while the children could play with their friends in the playroom! The younger sister could dance in the dance room lively, finish her homework at the third floor quietly or scrawl on the walls freely. This is a place where each child could find the interest and freedom, this is a place where the most wonderful childhood will leave in the deepest place of memory, this is a place where family love could tolerate the growth of every member, and this is a place where the children could not forget the wonderful years when they were young!

依景造城，共生共荣
BUILT ALONG THE SCENERIES WITH CO-EXISTENCE AND COMMON PROSPERITY

项目名称 ｜ 大连恒大檀溪郡23号
设计单位 ｜ 上海全筑建筑装饰设计有限公司
软装配套 ｜ 上海全品室内装饰配套工程有限公司
项目地点 ｜ 山东大连
项目面积 ｜ 350 ㎡

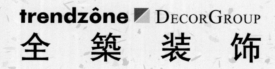

全 築 装 饰

上海全筑建筑装饰设计有限公司

Shanghai Trendzone Construction Decoration Design Co., Ltd

全筑装饰集团成立于1998年，是集建筑装饰研发与设计、施工、家具生产、装饰配套和建筑科技为一体的大型装饰集团。是中国建筑装饰协会常务理事单位和上海市装饰装修行业协会副会长单位，具有设计甲级和施工壹级资质。是中国驰名商标和上海市著名商标企业。蝉联十二届上海市室内设计大赛金奖，在中国装饰业界极具实力及美誉度。

全筑装饰集团拥有专业化部品加工基地及先进的生产设备，为装饰装修提供强大保障。同时建有专门装饰配套展厅，为装饰装修提供全面、专业化的后期配套服务。并在行业内率先成立了自主研发部门，为客户提供更为专业的系统服务。

　　大连恒大檀溪郡项目坐北朝南，背靠大连西郊森林公园，南向观海，循山而建，依景造城，实现人与自然的共生共荣。本案设计师以美式风格诠释，抒写多元、不羁的自由精神。素朴的色彩、奔放的材质。美式风格以宽大、舒适、杂糅各种风格而著称。在本案中，设计师摒弃了粗犷的一面，在古典的规则中增添了一点随意和巧思，使得兼顾品质生活和随性的业主获得一种妥帖的心灵安慰。整体设计传承经典，强化细节，给居者精致且经典的空间享受，在浓厚的空间气派中感受家庭的温馨与惬意。

　　Facing south with seascape, Evergrande Fores Town(Dalian) is located against Dalian West National Forest Park, built along the mountains and sceneries, realizing the co-existence and common prosperity between human and nature. The designer used American style to interpret the diversified and unruly spirit of freedom through simple colors and unrestrained materials. American style is famous for the broadness and comfort with a mix of different styles. In this case, the designer added a little leisure sense and ingenuity into the classical rules instead of roughness, making the owner who stressed on both qualified life and leisure achieve a proper comfort in mind. The overall design inherited the classics and intensified the details, giving the inhabitants a delicate yet classical enjoyment as well as the profound generousness in the space where they could feel the warmth and happiness of family.

轻蓝主义
SLIGHT BLUE-ISM

软装陈设 ｜杭州道境室内设计有限公司
项目地点 ｜江苏苏州
项目面积 ｜520 m²
主要材料 ｜古铜、大理石、桃花芯木等

杭州道境室内设计有限公司

道境，专业致力于房地产领域的室内设计公司，是优秀室内设计师共同工作的专业平台。

历经数载，道境已逐步发展成为百余人的工作团队，设计主创有着本专业相对丰厚的阅历和良好的职业素养；通过与境内外各专业领域团队不断的合作、交流，设计团队保持着设计前沿思想和创新活力；并通过与多家地产集团的真诚合作，道境设计已建立完善的设计项目全程监管体系，保证设计成果的完美呈现。

道境设计，希望通过一群有理想的设计团队不断努力，为客户提供更高品质的设计和服务。

杭州道境室内设计有限公司荣获2012-2013年度"全国室内装饰优秀设计机构"

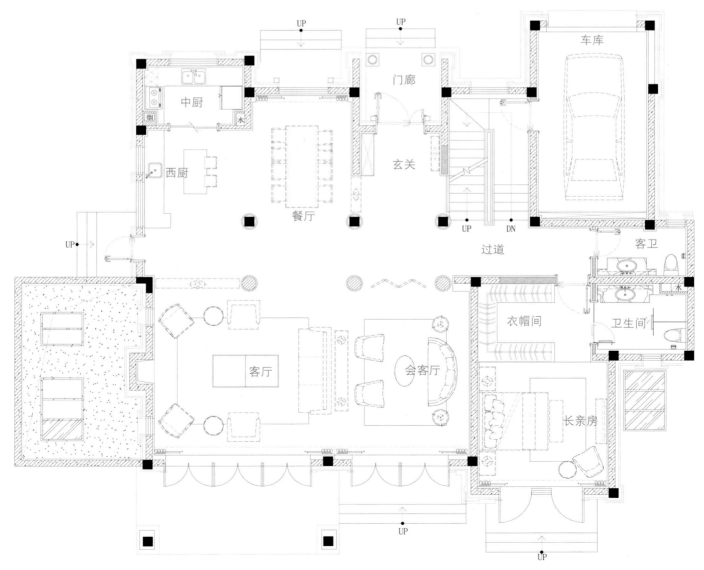

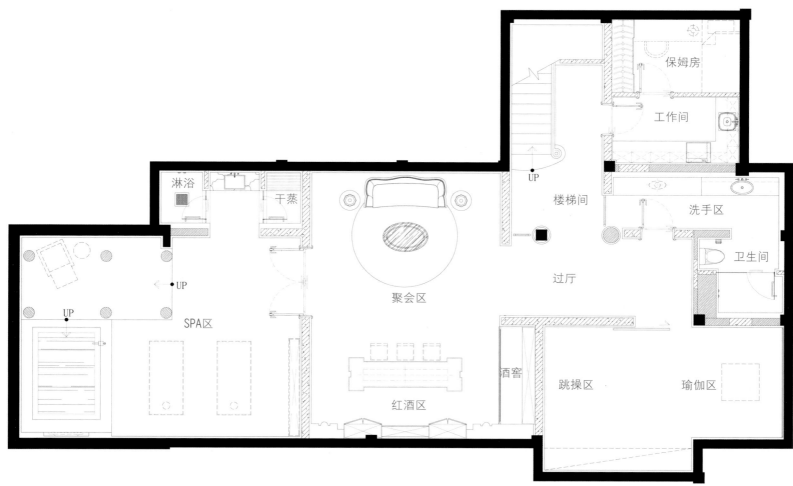

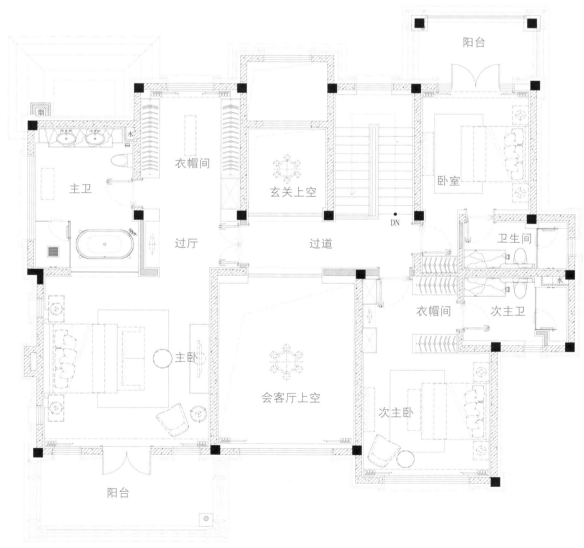

法式情缘
LOVE OF FRENCH STYLE

项目名称 ｜ 江阴红星美凯龙运河世家联排别墅
设计公司 ｜ 上海李孙建筑设计咨询有限公司
设计总监 ｜ 邢政
软装设计 ｜ 杨雯
项目地点 ｜ 江苏江阴
项目面积 ｜ 355 m²
主要材料 ｜ 实木、黑色高光漆、香槟金箔贴面等
摄 影 师 ｜ 陈盛

上海李孙建筑设计咨询有限公司

上海李孙建筑设计咨询有限公司是主要从事各类星级酒店、高级别墅、会所及餐饮娱乐空间等商业空间的室内装饰设计。正式成立于2007年，是一家集设计、创意一体的综合性企业。公司先有香港、台湾、马来西亚等境外及国内优秀设计师。他们以非凡的设计创意和精益求精的服务精神已在沪上成为具有一定影响力和知名度的公司团队。同时将继续以其严谨态度和创意设计为室内设计界打造完美的独一作品。

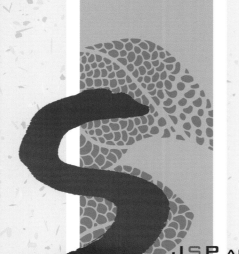

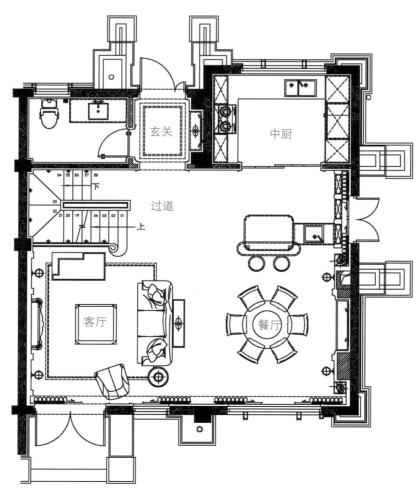

　　高贵而典雅是法式风格的代名词。整体色调以湖蓝色作为主调，为了让整体空间清新中透露着神秘，采用了橘红色作为点缀色。布局上突出轴线的对称，恢宏的气势，豪华舒适的居住空间。家具细节处理上运用了雕花、线条，制作工艺精细考究。点缀在自然中，崇尚冲突之美。不论是品鉴床头台灯图案中娇艳的花朵，抑或是品尝自己珍藏多年从酒窖刚拿出来的酒，或者听着留声机里传来清脆优雅的旋律，在任何一个角落，都能体会到主人悠然自得的生活和阳光般明媚的心情。

Nobility and elegance are the representatives of French style. Lake blue is used as the main tone in order to reveal the mysterious sense in the fresh air of the space, embellished by jacinth. The symmetry of the axis, the magnificent momentum and luxurious yet comfortable living space are highlighted on the layout. On the details of the furniture, the designer used carvings and lines delicately, which advocate the beauty of collision when embellished in nature. No matter the pattern of charming flowers on the table lamp, or the wine with many years' collection, even the elegant rhythm from the gramophone, one could feel the owner's leisure life and sunshine-like mood throughout the space.

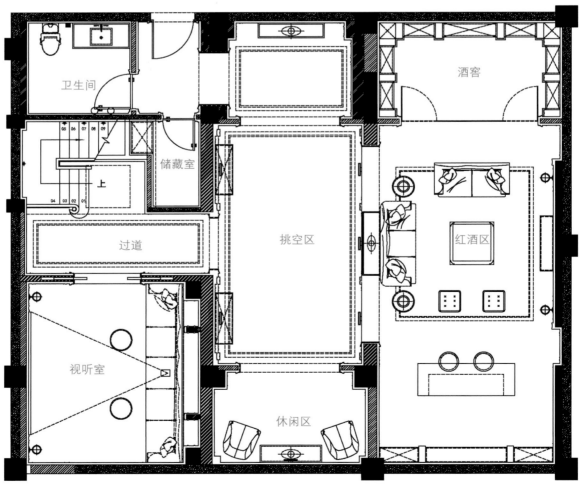

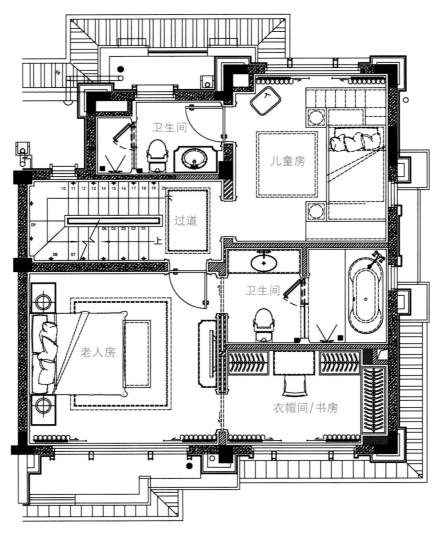

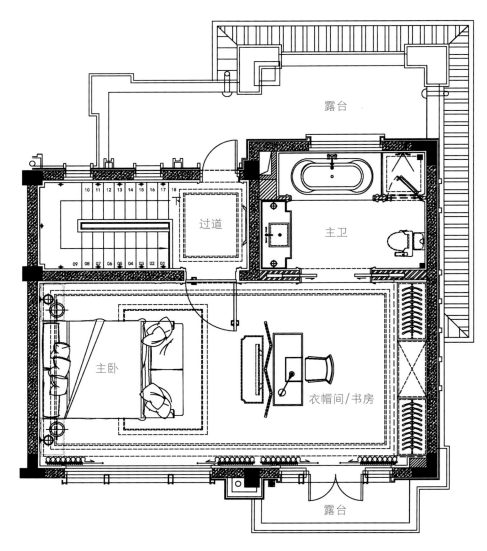

亚太豪宅大赏 III · 风靡大陆 | 245

复古英伦风
VINTAGE BRITISH STYLE

设计公司	杭州肯纳建筑景观设计有限公司
设 计 师	顾惠娟
项目地点	江苏吴江
项目面积	430 ㎡
主要材料	进口仿古砖、仿古樱桃木饰面、肌理涂料、墙纸等

顾惠娟 / TRACY

杭州肯纳建筑景观设计有限公司 设计总监

2004年，5位合伙人怀揣着设计理想踏上征程。秉持着对设计品质和建筑完美融合度的追求，最终获得了客户的关注与肯定。随着各种的设计项目的增加，我们组建了专业的景观设计团队、建筑设计团队、软装设计团队，KND成长为综合性服务机构。

11年，KND团队从4人到20人，由单一的室内设计团队到景观设计，软装设计等综合型的设计服务机构。自成立以来凭借自身团队细分的专业化优势先后完成的设计项目有：杭州西溪湿地小上海酒店、绍兴假日酒店、余姚管委会、余姚交通局、杭州市整形医院、杭州新天地K2地块等大型装饰设计项目。我们相信承载着不同情感的空间，更接近生活的本质，亦能感受更多浸染着不同文化和情绪的作品。

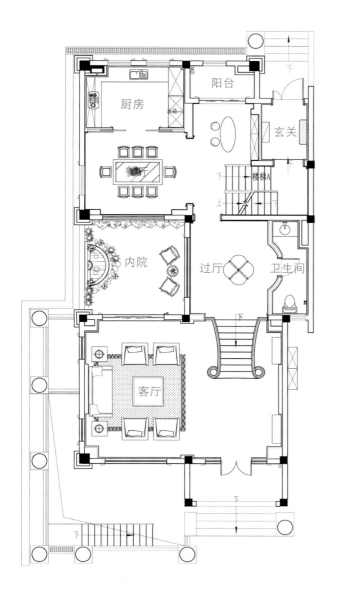

该案从硬装到软装围绕古朴而华丽的英伦气质展开设计。随着生活节奏的加快，传统的英式古典主义别墅设计风格往往显得过于厚重、繁复、缺乏变化，令人望而生畏，但同时又怀念其贵族气质，所以将古典与现代相结合，是现在最为稳妥的做法。

样板房为四层别墅，空间设计儒雅富丽，带有浓烈的英式色彩，设计师在咖啡色皮质沙发间放入了米色单人沙发，营造出明亮的视觉效果。地下一层的品酒区及书房彰显了主人的品味及生活情调，深红色的木饰面与地面简洁的大理石拼花融合，整个基调精雅又不失气度。同样古典英伦风格还讲究浪漫情怀与艺术感，所以自然少不了画家们的精美油画营造气氛，鲜花的运用更是给整个空间增添浪漫温馨的既视感。整个样板房带给我们复古的怀旧气息，又包含现代的时尚元素，这是最完美的邂逅。

The design of this case is started around the pristine yet gorgeous British temperament on both the hard and soft decoration. With the accelerated pace of life, traditional classical British villa design appears too heavy and complex without changes, which is awestruck and reminiscent of its noble temperament as well, thus combining classical and modern style together is the most proper method at present.

This is a show flat villa with four floors, while the space design of which is refined and gorgeous with strong British colors. The designer puts a single beige sofa into the leather sofa in coffee, creating a bright visual effect. The wine room and study at the basement manifest the taste and mood of the owner, while the dark red wooden veneer is combined with the concise marble on the ground, making the overall tone elegant with great presence. At the same time, classical British style emphasizes on the romantic feeling and artistic sense, so that there are a lot of delicate paintings, while the using of flowers adds romance and warmth into the whole space. The nostalgic flavor with the modern and fashionable elements throughout the show flat presents the most perfect encounter.

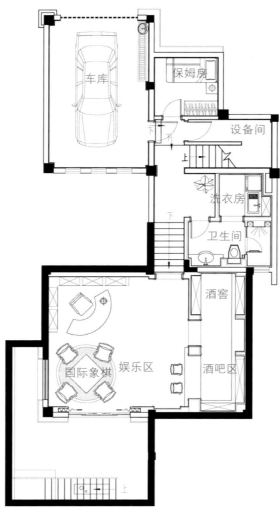

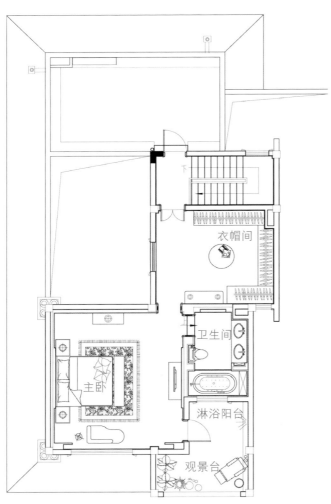

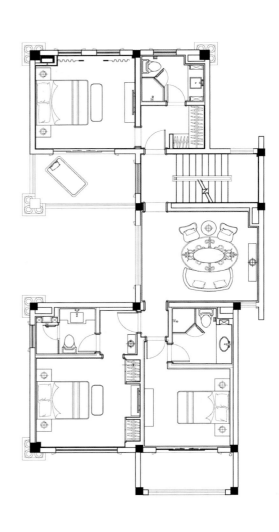

摩登轻法式
SLIGHT FRENCH STYLE IN FASHION

项目名称 ｜杭州翡翠城翠湖苑别墅样板房
设计公司 ｜西象建筑工程设计（上海）有限公司
设 计 师 ｜何文哲、唐瑶华
项目地点 ｜浙江杭州
项目面积 ｜654 ㎡
主要材料 ｜米黄、深灰、爵士白大理石，白色木饰面、竖纹木饰面、香槟金装饰铜条等

何文哲 WENZHE HE 西象建筑设计工程（上海）有限公司 总经理

何文哲，于2009年与一众心怀理想且富有激情的设计师们成立了上海迪扬建筑设计事务所，2012年成立西象建筑设计工程（上海）有限公司、西盛建筑设计（上海事务所）及何文哲（上海）设计咨询事务所。从事设计专业二十余年，孜孜不倦，致力于追求"空间与人物"的关系，创造"宜人宜心宜情宜居"的空间是我们的设计哲学。鲜明的设计概念需要具备丰富的想象力，我们不主张只采用一种独特的设计风格，需要运用不同的设计手法来诠释每一个项目，使之呈现别具一格的设计特点，同时确保设计成果的实用性、功能性以及所在时期和地域上的适用性。

唐瑶华 YAOHUA TANG 西象建筑设计工程（上海）有限公司 设计总监

唐瑶华，2004年毕业于上海工艺美术学院，现就读于同济大学EMBA、EMDN课程，从事设计行业十多余年，期间参与了众多工程项目，业务合作有绿城、金地、万科、中海、远洋、景瑞等大型房产公司。设计品味独到、有见地。

新兴的轻法式风格,优雅、温馨、悠闲,真真是花样年华,风情万种,传承优雅轻法式的古典唯美信仰。

本案的轻法式风格十分注重装饰性,但又不是传统的那种洛可可奢华装饰。精心选置世界各地的古董家具、装饰品,装饰画,把英国威廉时期、法国路易十四、路易十五时期的洛可可、巴洛克风格融入进整个空间,穿插其中当代时尚元素混搭出了不同的时尚风味。每一层都配有储物空间,让实用与美观兼具。

The new slight French style is elegant, warm and leisure, inheriting the classical and aesthetic belief of the elegant slight French style.

It emphasizes more on the decoration which is not the traditional luxury of Rococo style. Selected vintage furniture, decorations, decorative paintings from all over the world blend the Rococo and Baroque in William period of England as well as the stage of Louis XIV and XV of France into the whole space, while the mix of contemporary elements brings a different kind of fashion. Each floor is equipped with storing space, realizing both functions and beauty.

一下夹层到地下一层就能看到来自英国威廉时期的古董柜,可见男主人的生活品质。娱乐会客区域在采光上不是最理想,为解决这一问题,在材质、色调的运用上采用了对比度更强烈的色调。用水晶灯及大墙面装饰镜来提升整个空间的亮度,灰白的主色调,金、红点缀其中让人在视觉上感受到惊喜及冲击力。吧台的酒吧柜巧妙的利用了墙面空间,起到了装饰与储物兼具的功能。收藏室中来自欧洲的装饰挂画、轻中式风格的摆件,中西合璧的搭配给轻法式风格增添了别具风味的优雅。男主人从世界各地带回的古董家具、私人收藏陈设在雪茄吧,晤友会客时,惬意、随性。影音室的深色木饰、蓝黑色皮革、不锈钢包框的电影时尚海报,英伦细节穿插其中,让我们的视觉和听觉来一场前所未有的洗礼。

Going down the basement through the interlayer, you could see the vintage cabinet from the William period of England which presents the living quality of the host. The colors with strong contrast are adopted in the entertainment and reception area since the lighting here is not ideal. Chandeliers and huge decorative mirrors are used to enhance the lightness of the overall space, dominated by gray and white, while the embellishment of gold and red makes people feel surprised. Taken full advantage of the wall space ingeniously, the bar cabinet plays the functions of decoration and storage. The collocation of European decorative paintings and ornaments in slight Chinese style at the collection room adds a unique elegance into the slight French style. The vintage furniture and private furnishings brought from all over the world by the host are placed at the cigar bar, which is pleasant and leisure for meeting friends. The dark wooden ornaments, black blue leather, and film posters in stainless steel frames are placed throughout the media room, giving us a visual and audio feast unprecedentedly.a different kind of fashion.

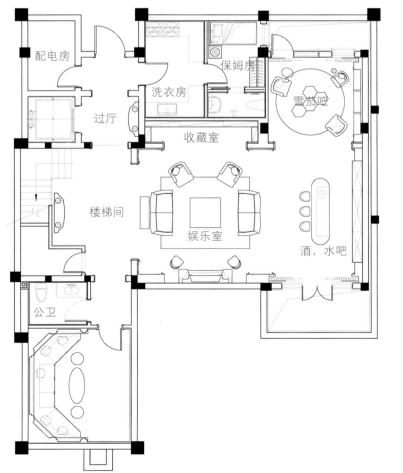
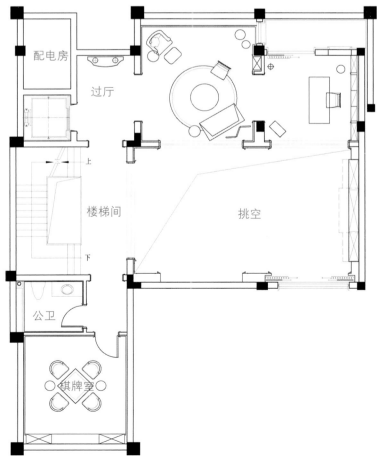

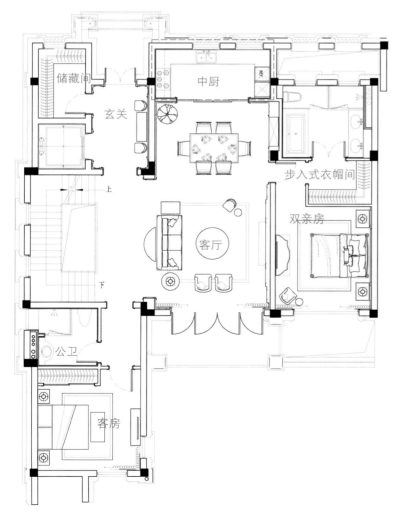

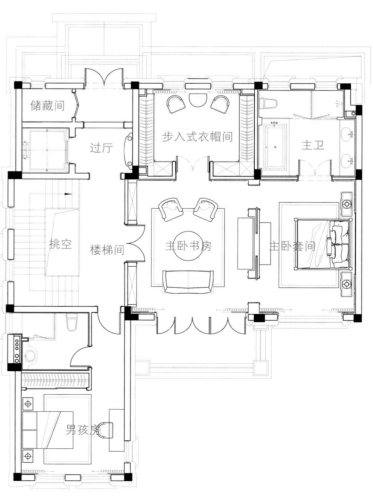

进门玄关处配置了兼具储物功能的衣帽间，深色的家具、点缀暗红色椅面，一进门就能让人感受到浓浓的生活气息。南北通的客餐厅可以直接步入外部的露天花园，让整个空间的采光及互动通透起来。餐厅的墙面摆放着各式餐盘，无关于繁华盛大，一点温情、一点品位、便足矣。父母房套间的色调蓝中点缀了些许的红，中式小摆件的运用，融入在整体风格中，亦是点睛之笔。温馨、舒适、雅致的空间更符合老年人的居住。客卧的窗户相对来说较小，在色彩的运用上橘色、天蓝提升了空间的明亮度，融于一层整体的装修风格。

The cloakroom with the function of storage is settled at the porch in the entrance, where the dark furniture and dark red chair surface make people feel a thick living atmosphere at once. One could walk into the outdoor garden through the transparent living room and dinning room, making the overall space full of light and ventilation. Varied plates are placed on the walls in the dinning room with a little warmth and taste. Red is embellished in the elders' suite with blue tone, while small Chinese ornaments that blended in the overall style are the highlights. A warm, comfortable and elegant space suits the elders better. The window in the guest room is relatively small, while orange and sky blue enhances the brightness of the space, which is blended into the overall decorative style at the first floor.

　　通往二层的踏步非常宽大，如履平地，在电梯厅和楼梯处同样也设置了玄关，直接连接了一步式阳台。步入主卧空间，延续了一层的轻法式风格，基调温馨的蓝绿色贯穿于整个二层空间，进门的小会客室，国外淘回的古董柜子及装饰画，无法复制的人文气息整个蔓延开来，和朋友、闺蜜在这里品着下午茶慢慢度过那阳光明媚的日子。主卧衣帽间加入了法式时尚元素，考究的细节、时尚而经典的配色宛若一位摩登女郎。卧房的软装饰非常温馨简洁，到处可见不同明度的绿色，大自然的色彩越窗而入。男孩房为乡村音乐风格的主题，小男主人兴趣爱好广泛，是个有些些小叛逆的少年。吉他、阅读、模型都是他的最爱，蓝色、橘色增添了活泼感。

The steps towards the second floor is very broad. There is another porch at the elevator hall and staircase, which is connected into the single-step balcony. Entering the master bedroom with the slight French style, where the warm blue and green tone are dominated. The vintage cabinets and decorative paintings collected abroad bring the nonreproducible humanistic flavor throughout the small reception room at the door, where you could spend your leisure time with close friends. French fashionable elements are added into the cloakroom in the master bedroom, where the fashionable and classical color collocation with refined details is just like a modern girl. The soft decorations here are very warm and concise with green color in different lightness. The boy's room is themed at country music, for the boy is a little traitorous with many hobbies and interests, such as guitar, reading and models. Blue and orange color add the vivacious sense here.

亚太豪宅大赏 Ⅲ · 风靡大陆 | 273

275

融入"历史与文化"中的低调奢华
LOW-KEY LUXURY BLENDED IN "HISTORY AND CULTURE"

项目名称	无锡酒店别墅样板房519户型
设计公司	DHA洪德成设计顾问(香港)有限公司
设 计 师	洪德成
项目地点	江苏无锡
项目面积	1102 m²
主要材料	乌斑木饰面、艺术夹丝玻璃、金属饰面、皮革布艺、特色墙纸、法国鎏金云石、幻彩玉石、贝克马赛克等

DHA国际设计联盟机构 创始人及董事长

洪德成精通摄影、绘画等艺术形式，常以独特的专业视角去发现生活中的被人们遗落的美，并通过设计去诠释和还原生活中的美，因而他的设计作品往往品味独特，不拘一格，让人耳目一新。这同样与他"设计不分国界，设计来源于生活"的理念密不可分。

DHA洪德成设计顾问（香港）有限公司，成立于2001年，拥有国际高端前沿视角，倡导设计、生活、艺术完美融合的理念，十余年来致力于为客户设计独一无二的高质量空间。DHA一向尊重客户意愿，充分发挥国际知名设计师团队的专业潜能，为客户量身定制国际化的专属设计。公司每年不定期邀请国际顶尖、不同文化背景且极具代表性的设计师在一起进行深度探讨交流，提升公司的国际化水平，同时推动设计行业的无界限发展。

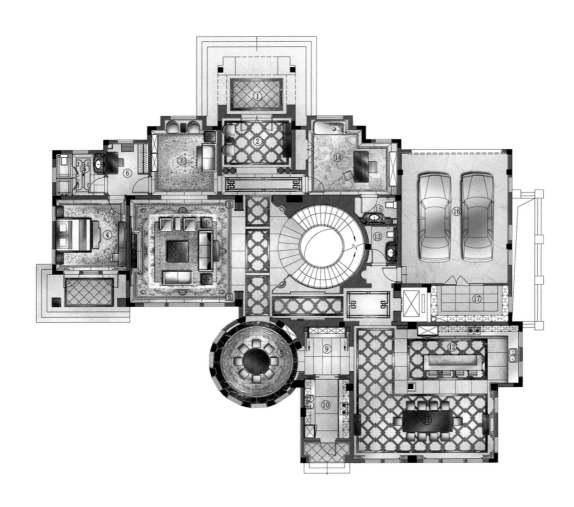

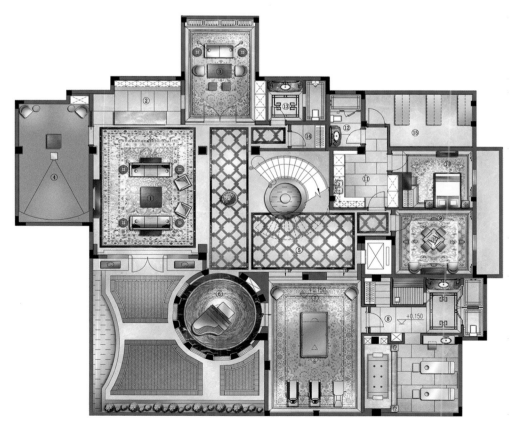

坐落于太湖翠峰叠嶂绿草如茵的人间天堂，与天水相隔一色的绝佳地段相结合，外观结构欧式建筑风格特点，设计师以法国路易十六文化历史设计为背景，营造出独具一格的高品质的低调奢华氛围。

室内空间设计体现设计师对于法式风格美感的独特表述，加上新颖独特的材质运用，整个空间的设计都透露出低调奢华的高品位气息。起居室玻璃金属质感红酒架延伸整体空间开阔视野，沉稳大气法式家具搭配精致带有异域风情艺术品点缀，营造出富有西方文化的品味与格调，凸显主人追求高雅的生活品质和文化素养，深厚的历史文化内涵凸显其中。一层最大亮点是吧台区，采用意大利天然纹样大理石作为背景。硬朗直线线条穿插于客厅整个空间中与法式古典家具、软装艺术品相互呼应，共同演绎着古典而又不失沉稳内涵的奢华空间。走廊过道的艺术雕塑小品为空间增加了动感的元素，地上铺的带有中世纪花卉图案的地毯，为整个空间增加了丰富的层次。

This case is located near the Tai lake which is a paradise with hills and plants around, and combined with the superexcellent section and European architectural features on the façade. The designer created a unique low-key luxurious atmosphere with high quality on the background of the culture and history of Louis XVI of France.

The interior design embodies the designer's unique expression on French beauty, added with novel and characteristic using of materials, making the design reveal the low-key luxurious flavor with good taste. The wine shelf with metal texture in the sitting room extended the broad view of the overall space, while the calm and generous French furniture was collocated with the embellishment of the artworks with delicate and exotic flavor, which created the taste and style with full of western culture and highlighted the owner's pursuit on elegant quality and cultural appreciation since the profound historical and cultural connotation were embodied. The highlight at the first floor is the bar where Italian marble with natural patterns were used as the background. Straight lines throughout the whole space echoed with Chinese and French classical furniture as well as soft decorations, interpreting the classical yet calm and connotive luxurious space. Small artful sculptures in the aisle added dynamic elements into the space, while the carpet with mediaeval patterns on the ground added rich layers into the whole space.

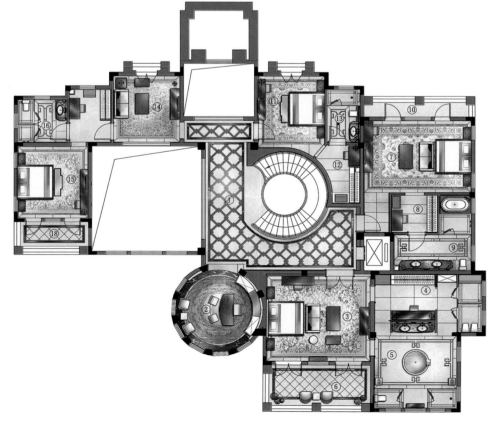

空间显得格外明亮而宽敞，华丽水晶吊顶星光熠熠，营造明亮舒适的整体环境。二层主卧墙面背景花纹图案，搭配极具夸张图案的花纹地毯绽放贵族的奢华气息，室内精美的家具在丝绸质感的床上用品的映衬下一点都不显得呆板，反而显得更尊贵和奢华。细节处的色彩搭配、材质对比层次更加体现了设计师独一无二的审美眼光。整体别墅空间融入现代简约法式设计手法，材质、色彩运用强调层次清晰而富有节奏韵律感。

The space appears bright and spacious, the gorgeous crystal ceiling of which was sparkling, creating a bright and comfortable environment. The patterns on the wall in the master bedroom at the second floor collocated with the patterned carpet, embodying noble and elegant sense, while the exquisite furniture appeared exalted and luxurious under the embellishment of the beddings with silk texture. The color collocation and comparison among materials on the details embodies the designer's unparalleled aesthetic vision further. Modern concise design techniques were blended into the whole space, where the using of materials and colors emphasized the layers with rhythms.

橄榄树之恋
LOVE OF THE OLIVE

项目名称 ｜ 林里 天怡样板间B户型
设计公司 ｜ 成象空间设计
软装公司 ｜ 成象空间设计
项目地点 ｜ 山东济南

成象空间设计公司 总经理

深圳成象空间设计公司 总经理

北京逸品成象空间设计公司 设计总监

　　岳蒙，用设计赞美生活，用修行完善生命，这是成象设计一直坚持的理念。对于成象空间设计的任何一位设计师而言，设计就是他们观察和改造这个世界的工具，他们每一个人都热爱这职业，他们用自己的专业给客户提供精确、职业、有创造力的设计为己任，努力给每一个项目带来超出客户预期的服务。以我所长，丈量世界之美。并且以自己的实力多次荣获各大奖项，其中2013、2014年金堂奖年度优秀样板间、售楼处设计；2014年"第九届金盘奖"总决赛年度最佳空间等。2015年中国室内设计年度评选金堂奖年度优秀样板间，2015年第十届金盘奖年度最佳样板房和年度媒体推荐奖等。

藏蓝色墨水的信写给穿白衬衫的他

悄悄掩在他的那本《瓦尔登湖》里

希望他看见，又愿他错过

黄昏时他在湖边低着头拨弄吉他

青草味的凉风替我拂过他的发

夕阳啊，你慢慢地沉

时间啊，你慢慢地淌

请给岁月以温柔，给我以阳光

在一汪湖水处

纵情好时光

The letter in dark blue ink was written for him in white shirt

Which was hidden quietly in his book called Walden

I hoped he could see it yet wish him missed it

He looked down and played the guitar by the lake when night fell

The breeze with the flavor of grass blew through his hair

I wished the sunset descended slowly as well as time

Please give the years gentleness and give me sunshine

With wonderful moments beside the water freely

静静坐着，数岁月的过往；微闭双眼，听湖水与风的吟唱。闲适是一种奢侈品，愈丰盛，愈安静。异国的一张明信片、旅途的机票……最终归于平静，万象掠过而不为所动。这里是我与自然共生的理想国。烤箱里，那人间烟火的味道，是马上烤好的芝士蛋糕。若找不到排解情绪的方法，就在床上假装思考。一盏暖灯，照亮了午夜梦回的心绪。是深邃蓝，还是宝石蓝？故事带着跳跃的灵感。一束阳光穿透云层洒向帐内，玩具恐龙和骑在小马上的娃娃在说悄悄话。宁静，是最美的人生之花。肤浅的人得不到它，浮躁的人无缘它。

Sitting quietly to recall the memories; closing eyes slightly to listen to the song of the water and wind. Leisure is a kind of luxury, which is more tranquil if it is more rich, such as a postcard from a foreign country and an air ticket of a trip… Everything calms down at last. This is the ideal land where I could live with nature. The cheese cake emits its fragrance from the oven. If you could not find the ways to untangle the emotions, you could stay on the bed to pretend thinking. A warm light lightens the mood at midnight. The story brings a jumping inspiration to make you wonder whether it is the deep blue or the sapphire? The sunshine sheds into the room through the clouds, while the dinosaur toy is whispering with the doll riding on the horse. Tranquility is the most beautiful flower in life, which could not be got by superficial and blundering people.

绿野仙踪
THE WIZARD OF OZ

项目名称 ｜ 贵阳乐湾别墅
设计公司 ｜ 广州道胜设计有限公司
主持设计师 ｜ 何永明
参与设计师 ｜ 道胜设计团队
项目地点 ｜ 贵州贵阳
项目面积 ｜ 248 ㎡
主要材料 ｜ 百合米黄大理石、凯撒灰大理石、雅士白大理石、黑白根大理石、复合木地板、白色钢琴漆饰面、香槟金拉丝不锈钢、扪皮、墙纸等
摄 影 师 ｜ 彭宇宪

广州道胜设计有限公司 创始人、设计总监

　　TONY HO于2005年成立广州道胜设计有限公司，设计项目分布在中国许多重要城市，以现代主义精神与热情为设计注入完美无瑕的风格和创新能量。透过整合建筑、室内设计、视觉图像和室内布置，每一次新作皆创造出独特的感官魅力与欢愉的空间气氛。历年不断获得国外50多个奖项，囊括CIID金奖、APIDA室内设计大奖……

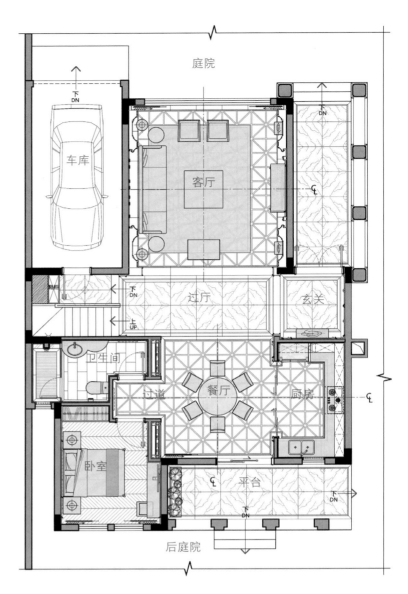

空间因人而生，因人而动，也因人而诞生有意义的形式。本案设计师运用格局与场域本身的特性，在空间方寸之中融入细腻的观察与思考。室内中以洗练纯净的白、高贵优雅的灰色为主色调，配以清新的绿色平衡出了无限舒适。

家具摒弃了繁复浮夸，配上适量的香槟金箔，同样勾勒出华贵气质。别具一格的铜质灯饰、精致独特的内饰，演绎空间的温馨与奢华，在浓烈的艺术氛围中体现浪漫自由的生活态度。

在茶室中这个富有蝉翼的独立空间中，糅合中西方艺术元素，透过丝丝禅意，体现业主超凡品味，呈现盎然的生活体验。

A space is formed and changed with meanings due to the existing of people. Designers in this case used the feature of the layout and the area itself to blend in exquisite observations and thinking. The clean and clear white and the noble and elegant gray were dominated in the interior, collocated with the fresh green, balancing the comfort anyway.

The furniture were out of complex and exaggerations, collocated with proper champagne gold foil, outlining the luxurious temperament. The peculiar bronze lamp and delicate accessories interpreted the warmth and luxury of the space, embodying the romantic and free attitude towards life in a strong artistic atmosphere.

In the private tea room, both Chinese and western artful elements were combined through a little Zen flavor, embodying the extraordinary taste of the owner and presenting the exuberant life experience.

素净雅致的主卧，半哑光素色真皮在深色木饰面的衬托下呈现丰富的层次感，精致的水晶饰品，也使得空间愈发沉稳内敛、精致优雅。淡淡黄绿色的点缀，使其个性鲜明、视觉感官强烈。

In the fresh and elegant master bedroom, the leather sofa presented a rich sense of layer under the embellishment of dark wood veneer, while the delicate crystal accessories made the space appear calm, restrained, delicate and elegant. The embellishment of yellow green brought a distinctive character and a strong visual sense.

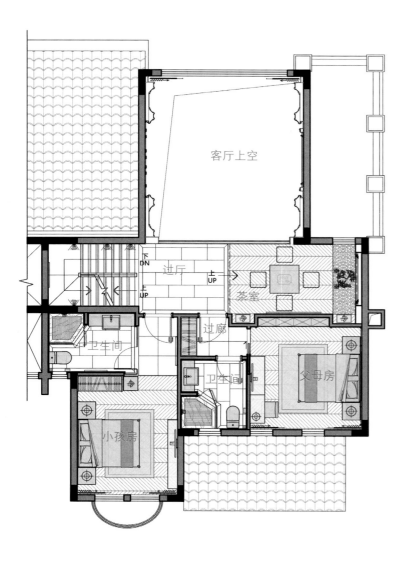

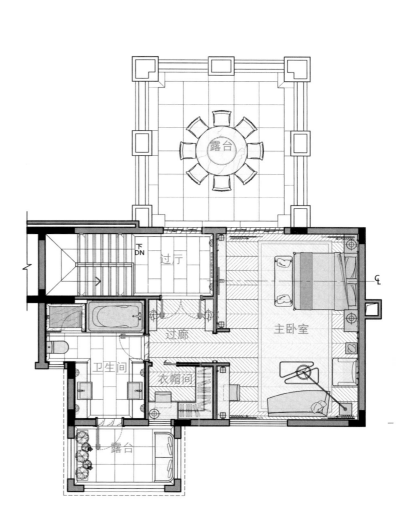

日暖绿城·典藏美宅
GREENTOWN WITH WARMTH·A MEMORABLE RESIDENCE

项目名称	绿城桃花源L户型样板房
设计公司	杭州尹泰瑞祺陈设艺术有限公司
软装设计	蔡东、应力玮、付雨鑫
项目地点	浙江杭州
项目面积	690 m²
主要材料	壁画、地毯、饰品、吊灯等

RICH 瑞祺
RICH Investment & Design

杭州尹泰瑞祺陈设艺术有限公司 Hangzhou Rich Display Art Co., Ltd

杭州尹泰瑞祺陈设艺术有限公司是一家集空间创意、软装设计与布场为一体的专业设计公司。拥有实力雄厚的设计团队，设计核心成员由具有国际视野的境外多位设计师担当。瑞祺设计对房产项目的市场定位、平面布局、空间规划、色彩搭配、成本控制和作业标准十分熟悉，擅长各路风格的售楼处、会所、样板房、高档酒店及写字楼的创意设计。代表作品有：绿城桃花源、绿城玉园、绿城之江一号、滨绿武林壹号、滨江华家池项目、联城水岸香榭、招商九龙仓雍景湾等。

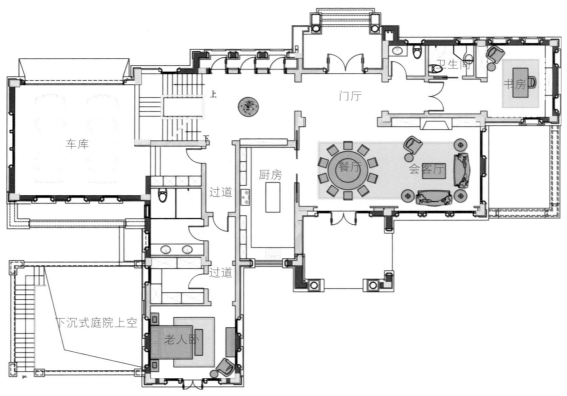

桃花源的天然资源、建筑设计及园林景观均为最上乘的配置，所以本案有着得天独厚的户外条件。设计师团队一直信奉少即是多的禅宗设计原则，更为重视房子的动线、光线及居住的品质感。室内采光及动线精妙合理，软装的搭配上谨守减法，空间里的每一处装点与陪衬都是相得益彰，不可或缺。因此设计师选用最具有代表性的品牌家具与饰品，融入大自然的配色，让建筑内外的景色交融。地下室的采光条件也是极好的，明亮通透，棋牌室旁的区域由于开间不足，将原本规划的台球室改为儿童娱乐室。设计师采用荷兰经典品牌Moooi及台湾精品动物Zuny的设计元素，规划了一个让小孩们爱不释手的主题游乐区，实为本案的一大亮点。

Based on the great configuration of the natural resources, architectural designs and garden views, this case owns the advantaged outdoor conditions. The design team, who believe in the design principle of Zen that less is more, and emphasize more on the generatrix, light and quality of the house. The lighting and generatrix indoor are delicate and reasonable, while the collocation on soft decoration is concise. Each decoration in the space complements each other quite well. Thus designers select the most representative brand furniture and accessories, blended with the colors in nature, making the sceneries both inside and outside the building integrated. The lighting of the basement is very good with brightness and transparency. The area beside the card room which has been planed to be a billiard room is changed into an entertainment room for the children since the lack of size. Designers adopt Moooi, a Dutch classical brand, and Zuny, the boutique animal from Taiwan to create a theme recreation area where children would like very much, which becomes a big highlight in this case.

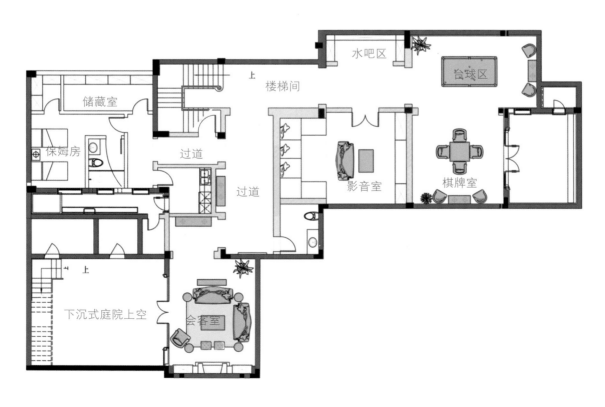

这里作为绿城17年来的首席别墅大盘，也是所有绿城别墅风格的发源地，包含法式、合院以及中式大宅等，有着太多的经典故事。设计师们在三个多月的时间内，每日尽心的研究绿城法式系的建筑及室内设计语言，意欲打造典藏美宅。

This has been the chief villa of Greentown since 17 years, which is also the origination of all styles within Greentown villas, including French villas, quadrangle dwellings and Chinese mansions with many classical stories. During more than three months, designers studied French buildings and interior design languages deliberately, willing to create a memorable and beautiful residence.

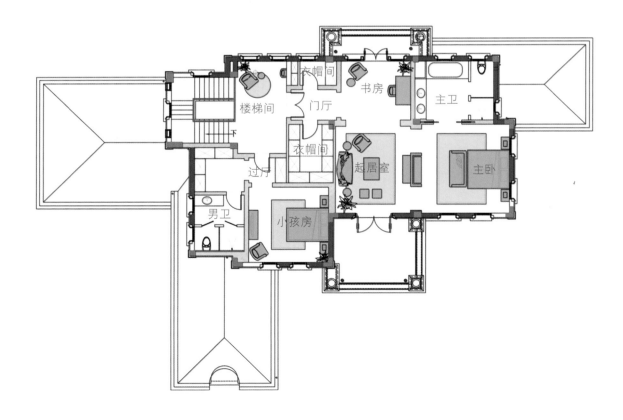

图书在版编目（CIP）数据

亚太豪宅大赏. 第3辑. 之风靡大陆 / 深圳视界文化传播有限公司编. -- 北京：中国林业出版社，2016.1
ISBN 978-7-5038-8378-1

Ⅰ. ①亚… Ⅱ. ①深… Ⅲ. ①室内装饰设计－亚太地区－图集 Ⅳ. ① TU238-64

中国版本图书馆CIP数据核字（2016）第020848号

编委会成员名单
策划制作：深圳视界文化传播有限公司（www.dvip-sz.com）
总 策 划：万绍东
责任编辑：蒙俊伶
装帧设计：张　梅
联系电话：0755-82834960

中国林业出版社　·　建筑家居出版分社
策　　划：纪　亮
责任编辑：纪　亮　王思源

出版：中国林业出版社
（100009 北京西城区德内大街刘海胡同 7 号）
http://lycb.forestry.gov.cn/
电话：（010）8314 3518
发行：中国林业出版社
印刷：深圳市汇亿丰印刷科技有限公司
版次：2016 年 1 月第 1 版
印次：2016 年 1 月第 1 次
开本：235mm×335mm，1/16
印张：20
字数：300 千字
定价：398.00 元 (USD 79.00)